Hubert Hahn

Higher Order Root-Locus Technique with Applications in Control System Design

Bernhard Herz

A Contribution about Controllability

Advances in Control Systems and Signal Processing
Editor: Irmfried Hartmann
Volume 2

Volume 1: Erhard Bühler and Dieter Franke
Topics in Identification and Distributed Parameter Systems

Volume 2: Hubert Hahn
Higher Order Root-Locus Technique with Applications in Control System Design
Bernhard Herz
A Contribution about Controllability

Hubert Hahn

Higher Order Root-Locus Technique with Applications in Control System Design

Bernhard Herz

A Contribution about Controllability

With 35 Figures

Friedr. Vieweg & Sohn Braunschweig/Wiesbaden

CIP-Kurztitelaufnahme der Deutschen Bibliothek

Hahn, Hubert:
Higher order root-locus technique with applications in control system design/Hubert Hahn. A contribution about controllability/Bernhard Herz. – Braunschweig, Wiesbaden: Vieweg, 1981.
(Advances in control systems and signal processing, Vol. 2)
ISBN 3-528-08490-1

NE: Herz, Bernhard: A contribution about controllability, gt

Editor:

Dr.-Ing. I. Hartmann
Prof. für Regelungstechnik und Systemdynamik
Technische Universität Berlin
Hardenbergstraße 29c
1000 Berlin 12, West Germany

Authors:

Dr. H. Hahn
Hornstr. 7
8000 München 40, West Germany

Dr. B. Herz
Prof. für Mathematik
Technische Universität Berlin
Fachbereich 3 – Mathematik
Straße des 17. Juni 135
1000 Berlin 12, West Germany

Druck und buchbinderische Verarbeitung:
IVD, Industrie u. Verlagsdruck, Walluf b. Wiesbaden
Printed in Germany

ISBN 3-528-08490-1

Preface

This bookseries aimes to report new developments on parts of control systems and signal processing at a high level, without a restriction in the size of contribution, if necessary. Basic knowledge is needed.

Each volume deals with a special theme. The following types of contributions will be taken into consideration:

- coherent representation on single topics with new results,
- new improved expositions or new applications in a known field of control systems and signal processing,
- practical applications of theoretic methods,
- a survey of published articles with deep insight.

Seminar work outs or themes from meetings with exeptional interest could belong to this series, if they satisfy the conditions.

The manuscript is reproduced by photographic process and therefore it must be typed with extreme care. The author can get instruction.

The manuscript is sent to Prof. Dr. I. Hartmann, Institut für Regelungstechnik, Technische Universität Berlin, Hardenbergstraße 29c, 1000 Berlin 12, West Germany.

Irmfried Hartmann

Berlin, January 1981

Contents

Higher Order Root-Locus Technique with Applications in Control System Design

Hubert Hahn

To Gerdi

Human nature being as it is, the very existence of science must be considered a miracle

Acknowledgment

The author would like to express his thanks to Prof. Dr. J. George, Math. Department, University of Wyoming, Laramie, USA, for many very useful discussions of the Newton Diagram Techniques during the years 1975 and 1976.The author received much valuable advice and scientific insight into the problems of Bifurcation Theory from this cooperation.

Thanks are also due to Prof.Dr. K. Kirchgässner, Mathematisches Institut A der Universität Stuttgart, W-Germany, for his interest in this work.

Improvements in the final draft were accomplished through the help of Prof. Dr. I. Hartmann,Institut für Regelungstechnik der Technischen Universität Berlin, W-Germany. His constructive criticism is greatly appreciated.

Finally, the author wants to express his gratitude to his wife Gerdi for typing the manuscript and for her understanding during the long hours of involvement with this book.

The author is happy to acknowledge the financial support of his research by the Deutsche Forschungsgemeinschaft (Contract No. Ha 977/1) in 1975 and 1976.

Also he thanks his employer Industrieanlagen-Betriebsgesellschaft Ottobrunn near Munich, and especially Mr. Raasch, for the permission to use the company's computers.

Contents

Notation and Symbols

Chapters are subdivided into sections, which are numbered I.1, I.2, ... , II.1, II.2, ... , and so on. Theorems, examples and figures are numbered consecutively within each chapter, prefixed by a chapter number. Tables are numbered with slight modifications. Tables a, b and c are related to the coefficients of the polynomials $Q_j(s)$ and $R_i(k)$, to the roots of $Q_j(s)$ and to the roots of $R_i(k)$, respectively. Table 5.b.2 , for example, contains the roots of the various polynomials $Q_j(s)$ related to the characteristic polynomial of Example 2 of Chapter V.

Operations

x^T	transpose of a vector x
$(x_1, \dots, x_n)^T$	column vector with components $x_1,\dots,x_n$
A^T	transpose of a matrix A
A^{-1}	inverse of a square regular matrix A
det A	determinant of a square matrix A
trace A	trace of a matrix A
$\begin{pmatrix} A, & B \\ C, & D \end{pmatrix}$	partitioning of a matrix A into blocks A,B,C and D.
$\mathrm{diag}(A_1, \dots, A_n)$	block diagonal matrix
$A > 0$, $A \geq 0$	matrix A is positive definite or non-negative definite, respectively
$\dot{x}$, dx(t)/dt	time derivative of the time-varying vector x(t)
Re{s}	real part of the complex number s
Im{s}	imaginary part of the complex number s
arg γ	argument of the complex number γ

$\min_i$	minimum with respect to i
$\max_i$	maximum with respect to i
$x \in X$	x is element of the set X
$X_i \subset X$	X_i is subset of the set X
$A = \bigcup_i A_i$	A is union set of the sets A_i

Symbols

a_j^i	coefficient of polynomials F(x,k),P(s,k)
A	plant matrix of a finite-dimensional linear differential system
B (or b)	input matrix (or input vector) of a finite-dimensional linear differential system
C (or c^T)	output matrix (or vector) of a finite-dimensional linear differential system
$\mathbb{C}^1$	complex numbers
e(s), e(k)	exponents of the variables s and k in the polynomial P(s,k)
e_I, e_{II}, e_{III}, ...	projections of the segments $\mathcal{L}_I$, $\mathcal{L}_{II}$, $\mathcal{L}_{III}$, ... of the exponent diagram onto the axis e(s) and e(k) , respectively
E	System matrix of the Euler-Lagrangian-Equations
F(x,k)	polynomial in the complex variable x and in the real parameter k
$G_c(s)$	closed-loop transfer function matrix
h(P)	height of the polynomial P(s,k)
I_n	n-dimensional identity matrix
$i = (-1)^{1/2}$	imaginary unit
k	real parameter,sometimes feedback factor

$K = k.K_1$	variable feedback matrix
K_1	constant feedback matrix
k_i^j	jth root of polynomial $R_i(k)$
$K^r\{k\}$	ring of power series of the form $\Sigma \gamma_j . \lambda^{j/r}$, $j \in \mathbb{N} \cup \{o\}$, $o < r \in \mathbb{N}$
$K^{\infty}\{r\} = \bigcup_{r \in \mathbb{N}} K^r\{k\}$	union of the rings $K^r\{k\}$
$K^{*}\{k\}$	ring of formal power series
$\mathcal{L}_j(P)$	jth segment of the exponent polygon related to $P(s,k)$
$\mathcal{L} = \bigcup_j \mathcal{L}_j(P)$	exponent polygon of $P(s,k)$
$l(\mathcal{L}_j)$, $l(\mathcal{L})$	length of $\mathcal{L}_j$ and of $\mathcal{L}$, respectively
$l_o(\mathcal{L})$	degree of degeneration of $\mathcal{L}$
$L(s)$	open-loop transfer function matrix
$M(s)$	closed-loop transfer function matrix
o , O	zero number, zero vector, zero matrix
$P(s,k)$	polynomial of the complex variable s and of the real parameter k
P	solution of the matrix Riccati Equation
Q	weighting matrix of the state vector x
$\mathbb{Q}$	rational numbers
$Q_j(s)$	subpolynomials of $P(s,k)$, related to factors k^j
$R(P,P_k)$	resultant of polynomials P and P_k
$\mathbb{R}^n$	real numbers
$R_j(k)$	subpolynomials of $P(s,k)$, related to factors s^j
r_i	exponents of the factor k in $P(s,k)$, $r_i \in Q$ or $r_i \in \mathbb{N}$

s_j^i	jth root of polynomial $Q_i(s)$
s_{oj}	jth asymptote point of the root locus asymptotes (for $k=\infty$)
$s(P)$	degree of $P(s,k)$ relative to s
$s(k)$	parameter-dependent root of $P(s,k)$
$u \in \mathbb{R}^p$	p-dimensional input vector of a system
U	real part of s
U_n	subspace of $\mathbb{R}^n$, or subset of $\mathbb{C}^1$
V	imaginary part of s
$x \in \mathbb{R}^n$	n-dimensional state vector
x_o	initial state vector
$y \in \mathbb{R}^p$	p-dimensional output vector
$\alpha_{I.1}, \dots$	slopes of segment I, ... of the exponent polygon $\mathcal{L}$
$\beta_{I.1}, \dots$	$\beta_{I.1} = \tan \alpha_{I.1}$, ..., where $\beta_I = \frac{\mu_I}{\nu_I}$
γ_j	jth root of the supporting polynomial $\phi(\gamma)$ of a segment of the exponent polygon $\mathcal{L}$
$\varkappa, \lambda, \mu, \nu, \sigma, \tau$	indices
$\varphi(o,o)$	angle of departure of a root locus branch from point $s=o$ for $k=o$
$\varphi(s_j^o, o)$	angle of departure from root s_j^o of polynomial $Q_o(s)$ for $k=o$
$\varphi(\infty,\infty)$	asymptote angle of a root locus branch for $k=\infty$
$\varphi(\infty, k_j)$	asymptote angle of a root locus branch for $k=k_j$

$\varphi(0,\infty)$	angle of arrival of a root locus branch at a point $s=0$ for $k=\infty$
$\varphi(s_j^q,\infty)$	angle of arrival of a root locus branch at a root s_j^q of $Q_q(s)$ for $k=\infty$
$\phi(\gamma)$	supporting polynomial of one of the segments $\mathcal{L}_j$,related to the exponent diagram
$\phi_A(s;k)$, $\psi_A(s; k)$	characteristic polynomial of matrix A
$\psi_c(s;k)$	characteristic polynomial of a linear closed-loop system

I. Introduction

1. About this book

The various methods and techniques used in linear control system analysis and design are usually classified by the attributes "classical" and "modern" .

Systems and plants that can be described in terms of classical control principles are linear and time-invariant, and have a single input and output. The designer's methods are a combination of analytical ones (e.g. Laplace transform , Routh test, ...), graphical ones (e.g. Bode plots, Nyquist plots, Nichols charts, ...), semi-analytical/semi-graphical ones (e.g. classical root-locus techniques, ...) with a good deal of empirical knowledge.
All of these methods are based on a few simple ideas of much intuitive appeal. And this is what practicing design engineers need. Unfortunately, to multiple-input systems, most of these methods are not directly applicable or far from being as effective as for single-input systems, and the designers ingenuity may even become an obstacle in achieving a satisfactory design.

Among the primary aims of modern ,as opposed to classical, control,are that of deempiricising control system design and that of presenting solutions for a much wider class of control problems than classical control can tackle.One of the most usual approaches modern control theory uses to achieve these aimes is that of providing an array of analytical design procedures to lessen the load of the task the designer's ingenuity is to accomplish and to shift more of the load to his mathematical ability and to the computational machines, that do the practical work.

The results of these computations are figures. Even if these figures are optimal in a certain sense, a lack of intuitive understanding is felt, a lack of insight into the qualitative behaviour of the system, when modern theory is compared to classical theory.

As a consequence,a series of activities have started in recent years, to combine the advantages of both, classical and modern control theory. Among other factors,the interrelationships and connections between these two approaches have been investigated in order to transfer concepts of classical control technique to multi-input systems and to include in modern control theory as many as possible of the simple ideas of classical theory (cf. [1.1], [1.2], [1.3], [1.4]).

The results presented in this monograph should be considered as one link in the chain of activities mentioned to connect classical and modern control theory. The book deals with the analysis and design of linear multi-input systems on the basis of <u>higher-order root locus technique</u>. This technique has been developed by the author in connection with investigations [1.5], [1.6] referring to the <u>singular element</u> of a nonlinear dynamic system as a main source of information relevant to both, the design of engineering applications and the understanding of certain biological phenomena (cf. [1.7], [1.8], and [1.9]). Most of the results presented in this monograph, may be found in Chapters III,IV,V,VI,IX and X of [1.6] .

The primary purpose of this book is to present an organized treatment of the higher-order root locus technique and its application to linear large scale and multi-input systems.

In <u>classical</u> root locus technique, the roots of a characteristic polynomial are constructed which depends <u>linearly</u> on a scalar parameter k. Linear single-loop systems with k as a feedback factor have a charateristic polynomial of this form.

Linear systems, in which more than one parameter is changed simultaneously, each of them being proportional to the other, usually have a characteristic polynomial of the form (1.1) which is <u>nonlinear</u> in a scalar parameter k.

$$P(s;k) = \sum_{i=0}^{q} \sum_{j=0}^{n_i} a_j^i . s^j . k^i \qquad (1.1)$$

Characteristic polynomials of this form by their nature appear in electric, mechanical, hydraulic, chemical and general thermodynamic systems and networks, whenever various elements are changed simultaneously and proportionally to each other, as e.g. in temperature-dependent systems or in the design of multi-degree-of-freedom hydraulic test facilities, in which the influence of a simultaneous variation of mechanical bearings, of hydraulic actuators, of servovalves and/or of feedback controllers of the various degrees of freedom is investigated.

Various approaches to construct the roots s(k) of equation (1.1) are known from literature (cf. [1.1o], [1.11], [1.12] and [1. 8]). They either treat very special cases of (1.1), are incomplete to construct the root locus plot, are too complicated to be useful in practice or are even incorrect. In this monograph, the roots s(k) of equation (1.1) are constructed by a set of rules, called <u>higher order root locus technique</u>. This technique has all the advantages of classical root locus technique and , at the same time, is applicable to multi-input systems with a characteristic polynomial of (1.1).

This technique is based on simple geometric ideas. It transforms the complex <u>algebraic structure</u> of the solution manifold of equation (1.1), with the various algebraic cases, into a <u>geometric pattern</u>, which is easy to draw. Therefore it can be easily applied to practical problems and appeals strongly to the intuitive and graphic trait of engineers. Moreover it will be shown to be a natural generalization of the classical root locus technique of Evans.

In connection with a digital computer, this technique provides a computer-aided design method, which gives insight into the system, allows to monitor the numerical results and indicates clearly how a given system design can be improved systematically.

The higher-order root locus technique has been applied by the author and his colleagues to complex industrial systems in the fields of spacecraft, aircraft dynamics, road vehicle dynamics and in the control of servohydraulic test facilities. Extremely simplified versions of some of these examples are presented.

A systematic application of a combination of higher-order root locus technique and describing function methods -abbreviated as "describing root locus technique"- to systems with several nonlinearities provides results that are in excellent agreement with experiments and with computer simulations. These results will be treated in another paper.

Computational aspects of the numerical difficulties involved in applying the higher order root locus technique to complex and large-scale systems -especially in connection with the computation of the parameter-dependent chracteristic polynomial (1.1) and the related roots- are omitted from this volume.

2. Chapter outline

The book is organized as follows.

Chapter II presents a short review of the Newton diagram technique for constructing all parameter-dependent "small solutions" of an algebraic equation of the form (1.1).

This technique is applied to two simple examples for illustration purposes.The mathematical foundation of the Newton diagram technique is presented in Appendices A.1 and A.2. The underlying theorems are reformulated and proved there.

The Newton diagram technique serves as a basis for the exponent diagram technique, developed by the author in [1.6].

This technique, derived in Chapter III, provides a graphic means for constructing all "small", "medium-scaled" and "large" parameter-dependent solutions s(k) of algebraic equations of the form (1.1).

This technique also serves as a basis for the higher-order root locus technique, derived in Chapter IV, for constructing the root locus plot of equation (1.1). Various construction rules are derived, most of which are easy to learn and simple to apply.

A variety of engineering examples from various fields of rigid body dynamics, from electrical network technique and from linear multi-loop control systems are treated by the higher-order root locus technique in Chapter V. For learning purposes,these examples are kept as simple as possible.

The higher-order root locus technique provides many simple synthesis rules and concrete hints for improving a linear multi-loop system. Related examples are presented in Chapter VI, where the classical and the higher-order root locus techniques are compared.

A systematic investigation of linear multi-loop control systems with characteristic polynomials of the form (1.1) is found in Chapter VII. Among other concepts, the general concept of a <u>sink</u> of a linear multi-loop control system is defined. The sinks are the finite goals of root locus branches with increasing k. They play the role of generalized zeroes within the root-locus concept.

The linear optimal control problem with a characteristic polynomial of the form (1.1) is discussed in Chapter VIII. It is shown that, in general, the asymptotes of a root locus plot of equation (1.1) do not constitute a Butterworth polynomial.

<u>Note</u>:

Those readers who are mainly interested in concrete practical applications should start by reading Chapters IV and V simultaneously.

References

[1.1] I.M. Horowitz, Synthesis of linear multivariable feedback control systems, Tr. IRE Aut.Control AC-5, pp.94-1o5, 196o.

[1.2] A.G.J. McFarlane, The retun-difference and return-ratio matrices and their use in the analysis and design of multivariable feedback control systems, Proc.IEE 117,pp. 2o37-2o49, 197o.

[1.3] H.H. Rosenbrock, Design of multivariable control systems using the inverse Nyquist array, Proc IEE,116, pp. 1929-1936 ,1969.

[1.4] C.H. Hsu, C.T. Chen, A proof of the stability of multivariable feedback systems, Proc. IEEE, 56, pp. 2o61-2o62, 1968.

[1.5] H. Hahn, Higher order root locus technique, a synthesis technique in singular system theory, Preprint, Department of Physics, University of Tübingen, W-Germany, 1975/76.

[1.6] H. Hahn, Zur Theorie und Technik singulärer Regelkreise, Habilitationsschrift,Fachbereich Physik, Universität Tübingen, 1977/78.

[1.7] H.Hahn, Geometrical Aspects of the Pseudo Steady State Hypothesis in Enzyme Reactions, Springer Lecture Notes in Biomathematics, Vol. 4, pp. 528-545, 1974.

[1.8] H.Hahn, The application of root locus technique to nonlinear control systems with multiple steady states, Int. J. Contr., Vol.27, No. 1, pp. 143-161 and 163-164, 1978.

[1.9] H.Hahn, Existence Criteria for Bifurcations as Stability Criteria for Critical Nonlinear Control Systems, Lecture Notes in Biomath. Vol. 21, pp. 3-49, 1978.

[1.1o] G.Rosenau, Höhere Wurzelortskurven bei Mehrgrössen-Regelsystemen, IFAC Symposium, Düsseldorf, 1968.

[1.11] P.G. Retallack, Extended root locus technique for design of linear multivariable feedback systems, Proc. IEE, Vol.117, No.3, 197o.

[1.12] J.J. Beletrutti, A.G.J. McFarlane, Characteristic loci technique in multivariable control system design, Proc. IEE, Vol.118, No.9, 1971.

II. Newton's Diagram

1. Statement of the problem

The Newton Polygon Technique provides the formal basis for the Exponent Diagram Technique and for the higher order root locus technique developed in the next chapters. A rigorous mathematical treatment of this technique is presented in the Appendix A of this monograph. Newton [21] considered the problem of finding <u>all</u> solutions of equation

$$f(x,k) = 0 \quad , \tag{2.1}$$

which tend to x_0 as $k \longrightarrow k_0$, when $f'_x(x_0,k_0) = 0$, and $f(x,k)$ can be expanded in positive integral powers of $(x-x_0)$ and $(k-k_0)$. As is well known from literature, the series solutions of the singular system (2.1) can not be written in the form

$$x = x_0 + \sum_{i=0}^{\infty} \gamma_i \cdot (k-k_0)^i, \qquad i \in \mathbb{N} \cup \{0\} \quad , \tag{2.2}$$

where the coefficients γ_i are determined recursively. There is a lack of information to start the appropriate recursive relations.

Newton looked for solutions of equation (2.1) in the form of the series

$$x = x_0 + \gamma_1 \cdot (k - k_0)^{\beta'_1} + \gamma_2 \cdot (k - k_0)^{\beta'_2} + \dots \quad , \tag{2.3}$$

where β'_1, β'_2, β'_3, ... is an increasing sequence of <u>rational</u> numbers. To determine the possible values γ_1, γ_2, ... and β'_1, β'_2, ... , Newton invented and employed a geometric device now known as Newton's Analytical Triangle or Newton's Diagram. Naturally the question of the convergence of such series was far beyond him. Further research, on the part of Lagrange, Puiseux and others showed that the fractional powers appearing in each series of type (2.3) have a finite common denominator and that these series converge in the neighbourhood of k_o (for $k_o \neq o$).

2. Construction of Newton's polygon

Rewriting equation (2.1) in the form

$$f(x',k') = \sum_{\substack{i=o \\ j=o}}^{\infty} a_j^i . x'^j . k'^i = o \,, \tag{2.4}$$

where $x' := x - x^o$, $k' := k - k_o$, Newton's diagram technique may be described as follows [2.2, 2.3, 2.4, 2.5, 2.6] :
We assume that there exist integers n and q such that $a_n^o \neq o$ and $a_o^q \neq o$, and use these letters for the minimal such.
We first plot, in the (j,i) - plane, the points for which $a_j^i \neq o$. Let $P_o = (n,o)$ and rotate the j-axis around P_o in the clockwise sense until it strikes some of the plotted lattice points in the first quadrant. Let $-\beta_1$ be the slope of the line obtained, and P_1 be the plotted point on this line with minimum j-coordinate. (cf. Figure 2.1). If P_1 lies on the i-axis we are through; if not, we continue this process obtaining a polygon with a finite number of sides $\mathcal{L}_1, \ldots ,$ $\mathcal{L}_g$. This polygon is well known as Newton's polygon.
Then the first term of the series expansion of the solutions (2.3) of the relation (2.4) takes the form

$$x = \gamma_1 \cdot k^{\beta_1} \,, \tag{2.5}$$

where $\beta_1 = \tan \alpha_1$, α_1 is taken from the Newton Diagram (cf. Figure 2.1), and γ_1 is a non zero root of the "supporting polynomial" corresponding to one of the sides $\mathcal{L}_1, \mathcal{L}_2, \ldots$ of the Newton polygon considered. The supporting polynomial of a side $\mathcal{L}_j$ is obtained :

(i) by omitting all terms of (2.4) which don't correspond to points on the straight line $\mathcal{L}_j$,

(ii) by replacing the factors x^j of (2.4) by means of factors γ_1^j , and

(iii) by omitting all factors k^i .

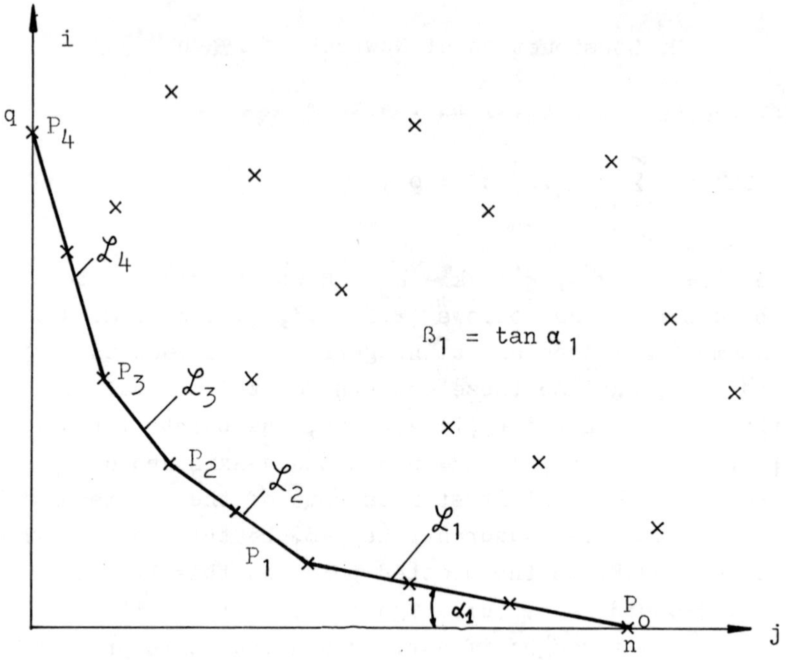

Figure 2.1: Newton Diagram

3. Examples of Newton's diagram technique.

The technique of Newton's diagram will now be used to compute all small solutions of two simple equations of the form (2.4) (compare [2.5] and [2.6]).

Example 1 :

Given the equation

$$F(x,k) = a.k^3.x + b.k^2.x^3 + c.k.x^5 = o \quad , \tag{2.6}$$

we are interested in all small solutions of (2.6) satisfying the relation

$$\lim_{k \to o} x(k) = o \quad . \tag{2.7}$$

The Newton diagram of (2.6) is drawn in Figure (2.2a) .

Case A : Let $a = 1$, $b = -2$, $c = 1$.

Ansatz of the small first order solutions of (2.6) :

$$x = \gamma . k^{\frac{1}{2}} \quad . \tag{2.8}$$

Supporting polynomial corresponding to (2.8) and to (2.6) :

$$\psi(\gamma) = \gamma - 2.\gamma^3 + \gamma^5 = \gamma.(\gamma^4 - 2.\gamma^2 + 1). \tag{2.9}$$

Roots of (2.9) :

$$\begin{aligned} \gamma_1 &= 0 \\ \gamma_2 = \gamma_3 &= 1 \\ \gamma_4 = \gamma_5 &= -1 \quad . \end{aligned} \tag{2.10}$$

First order branching solutions (small solutions) of (2.6) :

$$\begin{aligned} {}_1x &\equiv 0 \\ \left.\begin{aligned} {}_2x = {}_3x &= k^{\frac{1}{2}} \\ {}_4x = {}_5x &= -k^{\frac{1}{2}} \end{aligned}\right\} & \quad (k \geq 0). \end{aligned} \tag{2.11}$$

Ansatz of the small second order solutions of (2.6) :

$$x(k) = (1 + x_1).k^{\frac{1}{2}} \quad . \tag{2.12}$$

Insertion of (2.12) into (2.6) yields the relation :

$$\begin{aligned} {}_1F(x_1,k) = {} & k^{\frac{7}{2}} + k^{\frac{7}{2}}.x_1 - 2.k^2.(k^{\frac{3}{2}} + 3.k^{\frac{3}{2}}.x_1 + 3.k^{\frac{3}{2}}.x_1^2 + k^{\frac{3}{2}}.x_1^3) \\ & + k.(k^{\frac{5}{2}} + 5.k^{\frac{5}{2}}.x_1 + 10.k^{\frac{5}{2}}.x_1^2 + 10.k^{\frac{5}{2}}.x_1^3 + 5.k^{\frac{5}{2}}.x_1^4 + k^{\frac{5}{2}}.x_1^5) \end{aligned}$$

$$= k^{\frac{7}{2}}+k^{\frac{7}{2}}.x_1-2.k^{\frac{7}{2}}-6.k^{\frac{7}{2}}.x_1-6.k^{\frac{7}{2}}.x_1^2-2.k^{\frac{7}{2}}.x_1^3$$

$$+k^{\frac{7}{2}}+5.k^{\frac{7}{2}}.x_1+10.k^{\frac{7}{2}}.x_1^2+10.k^{\frac{7}{2}}.x_1^3+5.k^{\frac{7}{2}}.x_1^4+k^{\frac{7}{2}}.x_1^5 \qquad (2.13)$$

$$= 4.k^{\frac{7}{2}}.x_1^2+8.k^{\frac{7}{2}}.x_1^3+5.k^{\frac{7}{2}}.x_1^4+k^{\frac{7}{2}}.x_1^5 = 0 \quad .$$

As is shown by the Newton diagram corresponding to equation (2.13)(cf. Figure 2.2b), the relations (2.11) are already the exact small solutions of (2.6) (cf.Figure 2.2c).

Case B : Let $a = 8/9$, $b = -2$, $c = 1$.

Using again the ansatz (2.8) yields the supporting polynomial

$$\psi(\gamma) = \gamma.(\gamma^4 - 2.\gamma^2 + \tfrac{8}{9}) \quad , \qquad (2.14)$$

its roots

$$\gamma_1 = 0 \quad ,$$
$$\gamma_2 = +2.\sqrt{\tfrac{1}{3}} \;, \quad \gamma_4 = +\sqrt{\tfrac{2}{3}} \qquad (2.15)$$
$$\gamma_3 = -2.\sqrt{\tfrac{1}{3}} \;, \quad \gamma_5 = -\sqrt{\tfrac{2}{3}} \;,$$

and the first order small solutions

$$_1x = 0$$
$$_2x = +\tfrac{2}{\sqrt{3}}.k^{\frac{1}{2}}$$
$$_3x = -\tfrac{2}{\sqrt{3}}.k^{\frac{1}{2}} \qquad (\text{ for } \; k \geq 0) \qquad (2.16)$$
$$_4x = +\sqrt{\tfrac{2}{3}}.k^{\frac{1}{2}}$$
$$_5x = -\sqrt{\tfrac{2}{3}}.k^{\frac{1}{2}} \quad .$$

It is easily seen that the relations (2.16) are again the exact small solutions of (2.6) corresponding to case B (Fig. 2.2d).

Case C : Let $a = -1$, $b = -1$, $c = 1$.

In line with Cases A and B the small solutions of (2.6) take the form

$$
\begin{aligned}
{}_1x &\equiv 0 \\
{}_2x &= +\,k^{\frac{1}{2}}\cdot\left(\tfrac{1}{2} + \tfrac{1}{2}\cdot\sqrt{5}\,\right)^{\frac{1}{2}} \\
{}_3x &= -\,k^{\frac{1}{2}}\cdot\left(\tfrac{1}{2} + \tfrac{1}{2}\cdot\sqrt{5}\,\right)^{\frac{1}{2}}
\end{aligned}
\Bigg\}\; k \geq 0
$$

(cf. Figure 2.2e). (2.17)

$$
\begin{aligned}
{}_4x &= +\,k^{\frac{1}{2}}\cdot\left(\tfrac{1}{2} - \tfrac{1}{2}\cdot\sqrt{5}\,\right)^{\frac{1}{2}} \\
{}_5x &= -\,k^{\frac{1}{2}}\cdot\left(\tfrac{1}{2} - \tfrac{1}{2}\cdot\sqrt{5}\,\right)^{\frac{1}{2}}
\end{aligned}
\Bigg\}\; k \leq 0
$$

Case D : Let $a = 1$, $b = -1$, $c = 1$.

The trivial solution $x(k) \equiv 0$ is the only real solution of equation (2.6) corresponding to Case D apart from the trivial bifurcating solution.

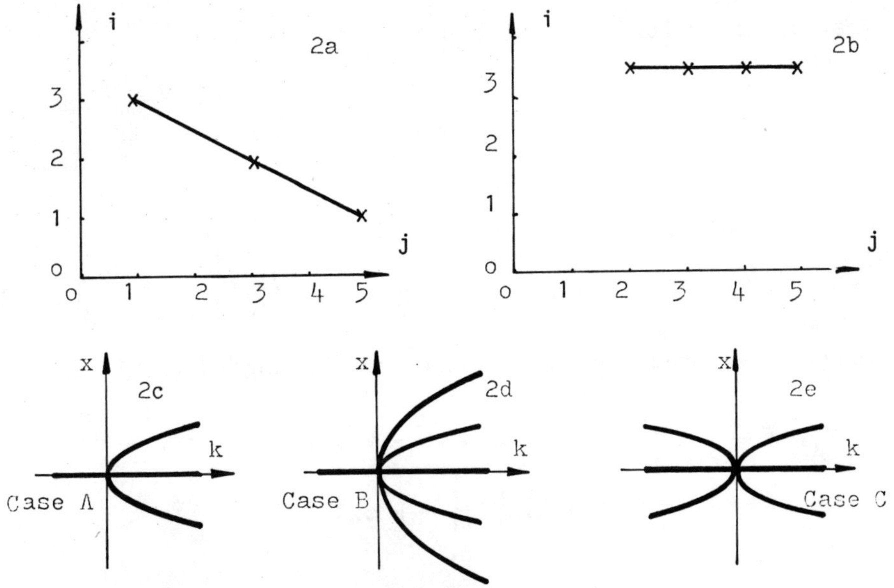

Figure 2.2: Newton and branching diagrams related to solutions of equation (2.6), Cases A,B and C .

Example 2 :

Given the equation

$$F(x,k) = k^3.x - k.x^2 + x^4 = 0 . \qquad (2.18)$$

The Newton diagram corresponding to (2.18) is shown in Figure 2. 3a .

Case A : $\beta_1 = 2$

Ansatz of the first order small solution :

$$x = \gamma.k^2 \quad . \qquad (2.19)$$

Supporting polynomial :

$$\psi(\gamma) = \gamma.(1 - \gamma) \qquad (2.20)$$

Roots of (2.2o) :

$$\gamma_1 = 0$$
$$\gamma_2 = 1 \quad . \qquad (2.21)$$

First order branching solutions :

$${}_1x = 0$$
$${}_2x = k^2 \quad . \qquad (2.22)$$

Ansatz of the second order small solutions

$$x = k^2.(1 + x_1) . \qquad (2.23)$$

Second order branching equation corresponding to (2.23) :

$${}_1F(x_1,k) = k^5 + k^5.x_1 - k^5 - 2.k^5.x_1 - k^5.x_1^2 + k^8 + 4.k^8.x_1$$
$$+ 6.k^8.x_1^2 + 4.k^8.x_1^3 + k^8.x_1^4$$

or

$$ {}_1F(x_1,k) = k^8 + (-k^5 + 4.k^8).x_1 + (-k^5 + 6.k^8).x_1^2 $$
$$ + 4.k^8.x_1^3 + k^8.x_1^4 = 0 \quad . \tag{2.24} $$

Using the Newton diagram of Figure 2.3b results in the second order ansatz

$$ x_1 = \gamma.k^3 \quad , \tag{2.25} $$

the supporting polynomial

$$ \psi(\gamma) = 1 - \gamma \quad , \tag{2.26} $$

its root

$$ \gamma = 1 \tag{2.27} $$

and the second order small solution

$$ x_1 = k^3 \quad . \tag{2.28} $$

The complete first and second order solutions take the form

$$ {}_1x = 0 $$
$$ {}_2x = k^2 + k^5 \quad . \tag{2.29} $$

Ansatz of the third order small solutions

$$ x = k^2 + k^5 + k^5.x_2 \quad . \tag{2.30} $$

Third order branching equation corresponding to (2.3o) :

$$ {}_2F(x_2,k) = k^5+k^8+k^8.x_2-k^5-k^{11}-k^{11}.x_2^2-2.k^8 $$
$$ - 2.k^8.x_2-2.k^{11}.x_2+k^8+k^{20}+k^{20}.x_2^4+4.k^{11} $$
$$ + 4.k^{11}.x_2+4.k^{17}+4.k^{20}.x_2+4.k^{17}.x_2^3+4.k^{20}.x_2^3 $$
$$ + 6.k^{14}+6.k^{14}.x_2^2+6.k^{20}.x_2^2+12.k^{14}.x_2 $$
$$ + 12.k^{17}.x_2+12.k^{17}.x_2^2 $$

or

$$ {}_{2}F(x_2,k) = 3.k^{11}+6.k^{14}+4.k^{17}+k^{20}+ (-k^8+2.k^{11}+12.k^{14} $$
$$ + 12.k^{17}+4.k^{20}).x_2+(-k^{11}+6.k^{14}+12.k^{17}+6.k^{20}).x_2^2 $$
$$ + (4.k^{17}+4.k^{20}).x_2^3+k^{20}.x_2^4 = 0 \quad . \tag{2.31} $$

Using the Newton diagram of Figure 2.3c results in the third order ansatz

$$ x_2 = \gamma .k^3 \quad , \tag{2.32} $$

the supporting polynomial

$$ \psi(\gamma) = 3 - \gamma \quad , \tag{2.33} $$

its root $\gamma = 3$ (2.34)

and the third order small solution

$$ x_2 = 3.k^3 \quad . \tag{2.35} $$

The complete first, second and third order solution takes the form

$$ {}_{2}x = k^2 + k^5 + 3.k^8 \quad . \tag{2.36} $$

Case B : $\beta_2 = 1/2$

Ansatz of the first order small solution :

$$ x = \gamma .k^{\frac{1}{2}} \quad . \tag{2.37} $$

Supporting polynomial :

$$ \psi(\gamma) = \gamma^2.(\gamma^2 - 1) \quad . \tag{2.38} $$

Roots of (2.38)

$$ \gamma_1 = 0 $$
$$ \gamma_2 = 1 $$
$$ \gamma_3 = -1 \quad . \tag{2.39} $$

First order small solutions :

$$ {}_1x = 0 $$
$$ {}_2x = k^{\frac{1}{2}} \quad (\text{for} \quad k \geq 0) \tag{2.40} $$
$$ {}_3x = -k^{\frac{1}{2}} \quad . $$

Ansatz of the second order small solutions :

$$ x = k^{\frac{1}{2}} + k^{\frac{1}{2}}.x_1 \quad . \tag{2.41} $$

Second order branching equation corresponding to (2.41) :

$$ {}_1F(x,k) = k^{\frac{7}{2}}+k^{\frac{7}{2}}.x_1-k^2-2.k^2.x_1-k^2.x_1^2+k^2+4.k^2.x_1 \tag{2.42} $$
$$ +6.k^2.x_1^2+4.k^2.x_1^3+4.k^2.x_1^4 $$
$$ = k^{\frac{7}{2}}+(2.k^2+k^{\frac{7}{2}}).x_1+5.k^2.x_1^2+4.k^2.x_1^3+4.k^2.x_1^4 = 0 \quad . $$

Using the Newton diagram of Figure 2.3d results in the second order ansatz

$$ x_1 = \gamma.k^{\frac{3}{2}} \quad , \tag{2.43} $$

the supporting polynomial

$$ \psi(\gamma) = 1 + 2.\gamma \quad , \tag{2.44} $$

its root

$$ \gamma_1 = -\frac{1}{2} \tag{2.45} $$

and the second order small solution

$$ x_1 = -\frac{1}{2}.k^{\frac{3}{2}} \quad . \tag{2.46} $$

The complete first and second order solutions take the form

$$_2x(k) = k^{\frac{1}{2}} - \frac{1}{2}.k^2 \quad . \tag{2.47}$$

Ansatz of the third order small solutions

$$x = -\,k^{\frac{1}{2}} + k^{\frac{1}{2}}.x_1 \quad . \tag{2.48}$$

Third order branching solution corresponding to (2.48) :

$$\begin{aligned} {}_1F(x_1,k) &= -\,k^{\frac{7}{2}}+k^{\frac{7}{2}}.x_1-k^2+2.k^2.x_1-k^2.x_1^2+k^2-4.k^2.x_1 \\ &\quad + 6.k^2.x_1^2-4.k^2.x_1^3+k^2.x_1^4 \\ &= -\,k^{\frac{7}{2}}+(-2.k^2+k^{\frac{7}{2}}).x_1+5.k^2.x_1^2-4.k^2.x_1^3+k^2.x_1^4 = 0 \quad . \end{aligned} \tag{2.49}$$

Using the Newton diagram of Figure 2.3e results in line with the solution (2.36) :

$$x_1 = -\,\frac{1}{2}.k^{\frac{3}{2}} \tag{2.50}$$

and

$$_3x = -\,k^{\frac{1}{2}} - \frac{1}{2}.k^2 \text{ (for } k \geq 0) \quad . \tag{2.51}$$

As a result, the small solutions of example 2 take the form (cf. Figure 2.3f) :

$$\begin{aligned} {}_1x &= 0 \\ {}_2x &= k^2 + k^5 + 3.k^8 + o(k^8) \\ {}_3x &= k^{\frac{1}{2}} - \frac{1}{2}.k^2 + o(k^2) \\ {}_4x &= -k^{\frac{1}{2}} - \frac{1}{2}.k^2 + o(k^2) \end{aligned} \quad \left.\begin{matrix} \\ \\ \end{matrix}\right\} (k \geq 0) \quad . \tag{2.52}$$

These solutions are only <u>approximate</u> small solutions of (2.18).

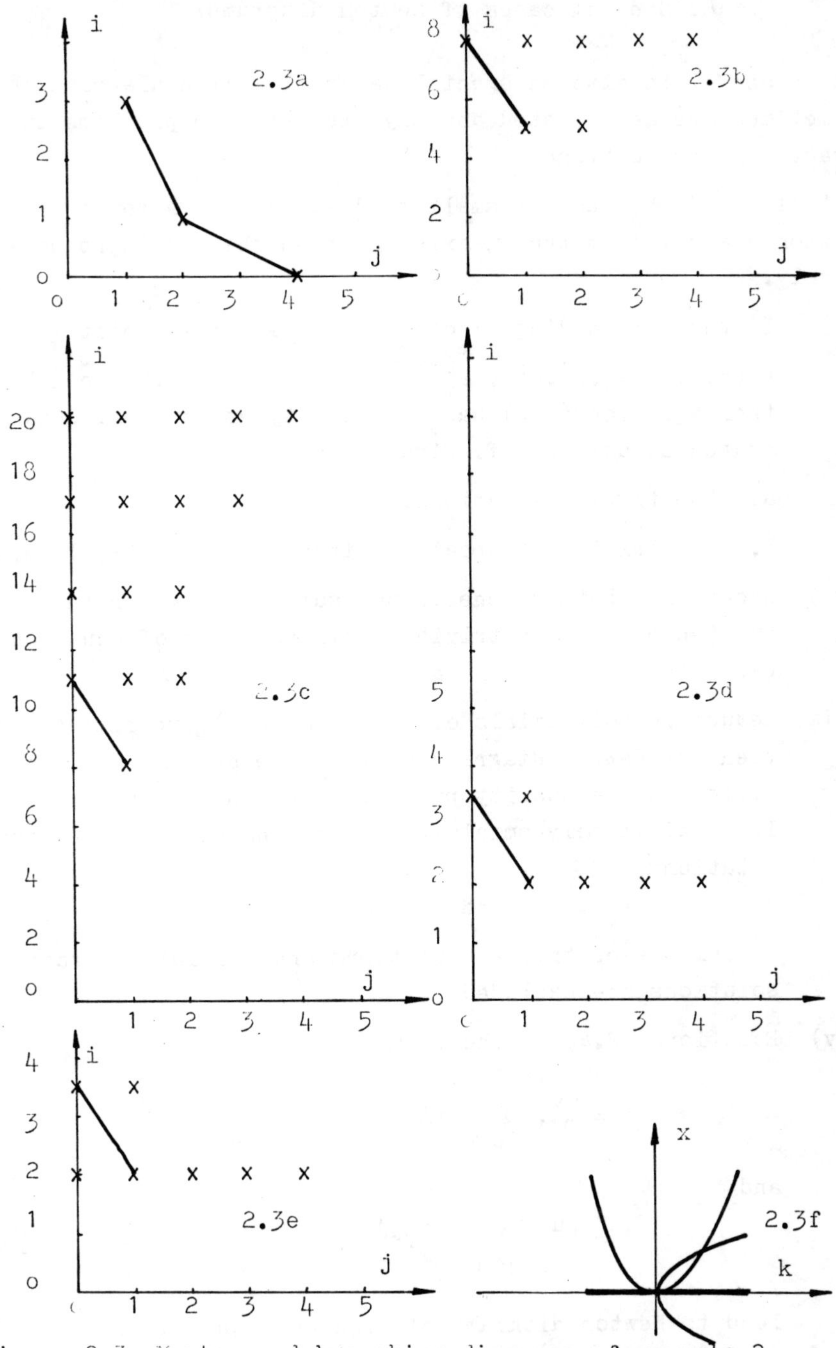

Figure 2.3: Newton and branching diagrams of example 2 .

4. Special cases of Newton diagrams

It is useful to discuss special cases of Newton diagrams which sometimes are degenerated and may give rise to problems in practical applications .

Let l_o and h_o be the smallest distances of a point of the Newton diagram from the i-axis and from the j-axis,respectively.

(i) If the Newton diagram consists of a single point ($f(x,k) = a_\nu^\mu . x^\nu . k^\mu$, $a_\nu^\mu \neq 0$; cf. Figures 2.4a to 2.4c), then equation (2.4) has only two types of small degenerated solutions (cf. Figure 2.4d) :

a. the trivial solutions,

b. the trivial bifurcations from the point $(x,k)=(0,0)$.

(ii) Necessary (but in general not sufficient) for the existence of a non trivial small solution of equation (2.4) is the relation $a_o^o = 0$.

(iii) Reducible polynomials of the form (2.4) give rise to more then one Newton diagram (compare the separation principle and the duality principle in control theory). Irreducible polynomials (2.4) are characterized by the relations

$$l_o = 0 \quad \text{and} \quad h_o = 0 \quad . \qquad (2.53)$$

In this case, trivial solutions and trivial bifurcating solutions are excluded .

(iv) Equations (2.4) of the form

$$f = x^\nu . \sum_{i \in \mathbb{N}} a_\nu^i . k^i \qquad (2.54a)$$

and

$$f = k^\mu . \sum_{j \in \mathbb{N}} a_j^\mu . x^j \qquad (2.54b)$$

lead to Newton diagrams of Figures 2.4e and 2.4f, respectively.

(v) Equations (2.4) of the form

$$f = \sum_{l=0,1,\dots} a_{i_l}^{j_l} x^{i_l} k^{j_l} \quad , \qquad (2.55)$$

where $i_\varkappa < i_\lambda$ and $j_\varkappa < j_\lambda$ for $\varkappa < \lambda$ have no small solutions of the form (2.5)(cf. Figure 2.4g).

(vi) The Newton diagram corresponding to a characteristic polynomial P(s;k) of a linear single loop control system with feedback factor k is shown in Figure 2.4h with x replaced by s (Compare Chapter IV.2, Step 1). Small solutions s(k) of P(s;k) exist only if the transfer function of the system has a vanishing pole (which is not compensated by a vanishing zero).

(vii) Let a segment of the Newton polygon meet exactly two points of the Newton diagram (cf. Figures 2.4i to 2.4l). Then (omitting the trivial solutions and bifurcations) the supporting polynomial corresponding to this segment takes the form

$$b.k^{\mu} + g.x^{\nu} = 0 \quad , \qquad (2.56)$$

and the <u>real</u> small solutions of (2.4) are among those diagrams in Figures 2.5a and 2.5b.

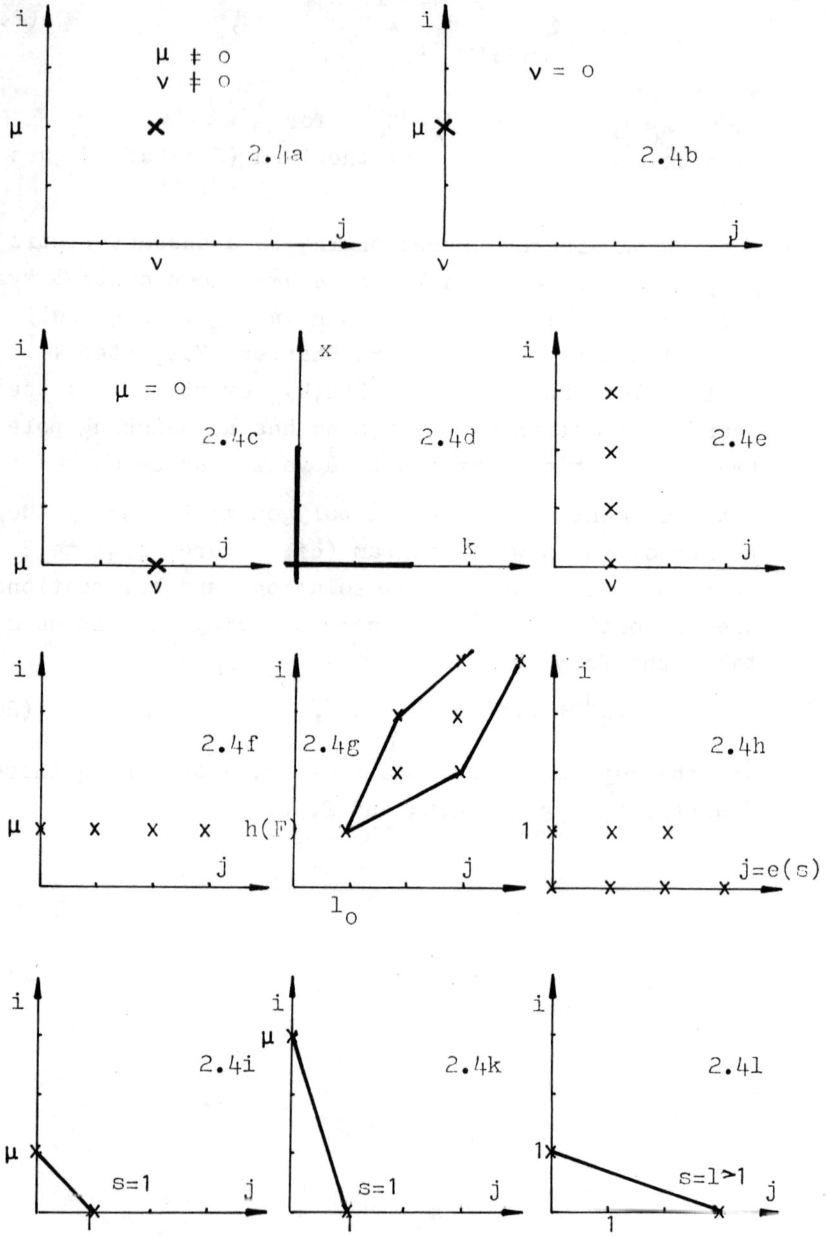

Figure 2.4: Simple Newton and branching diagrams

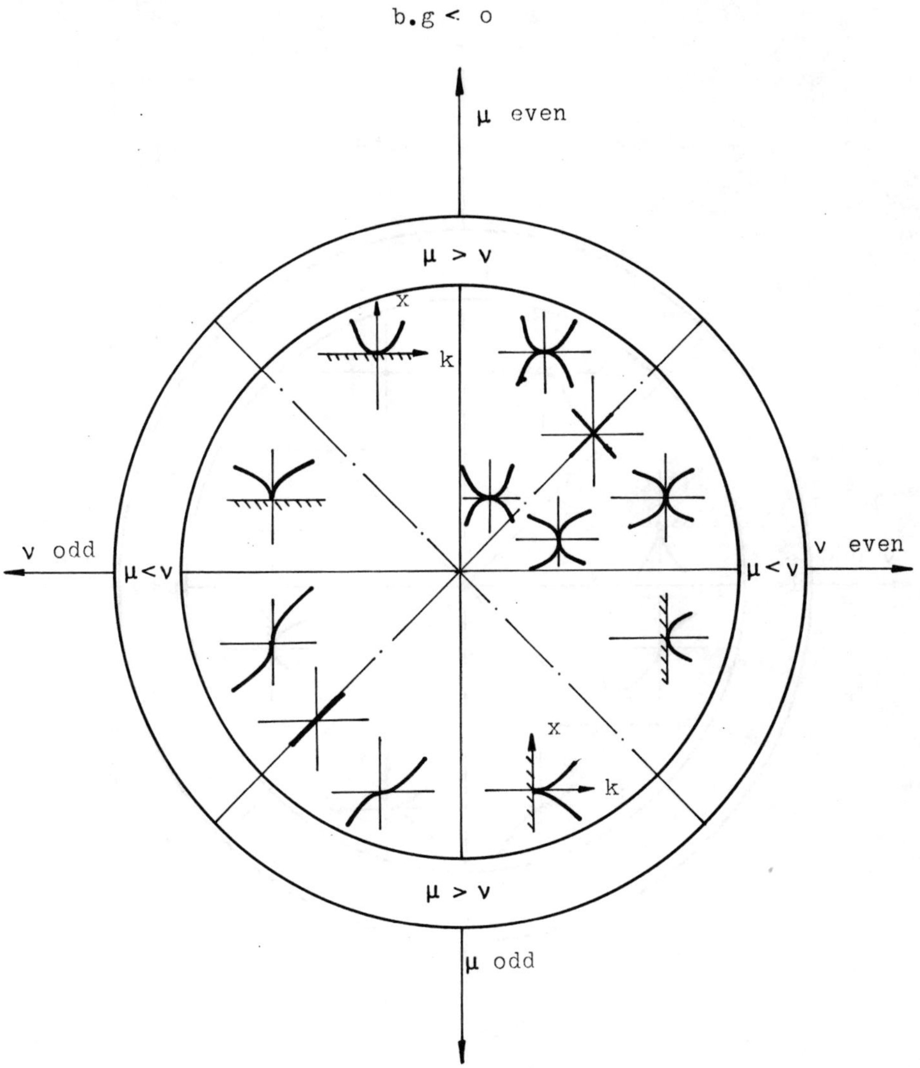

Figure 2.5.a: Elementary real branching diagrams of (2.56).

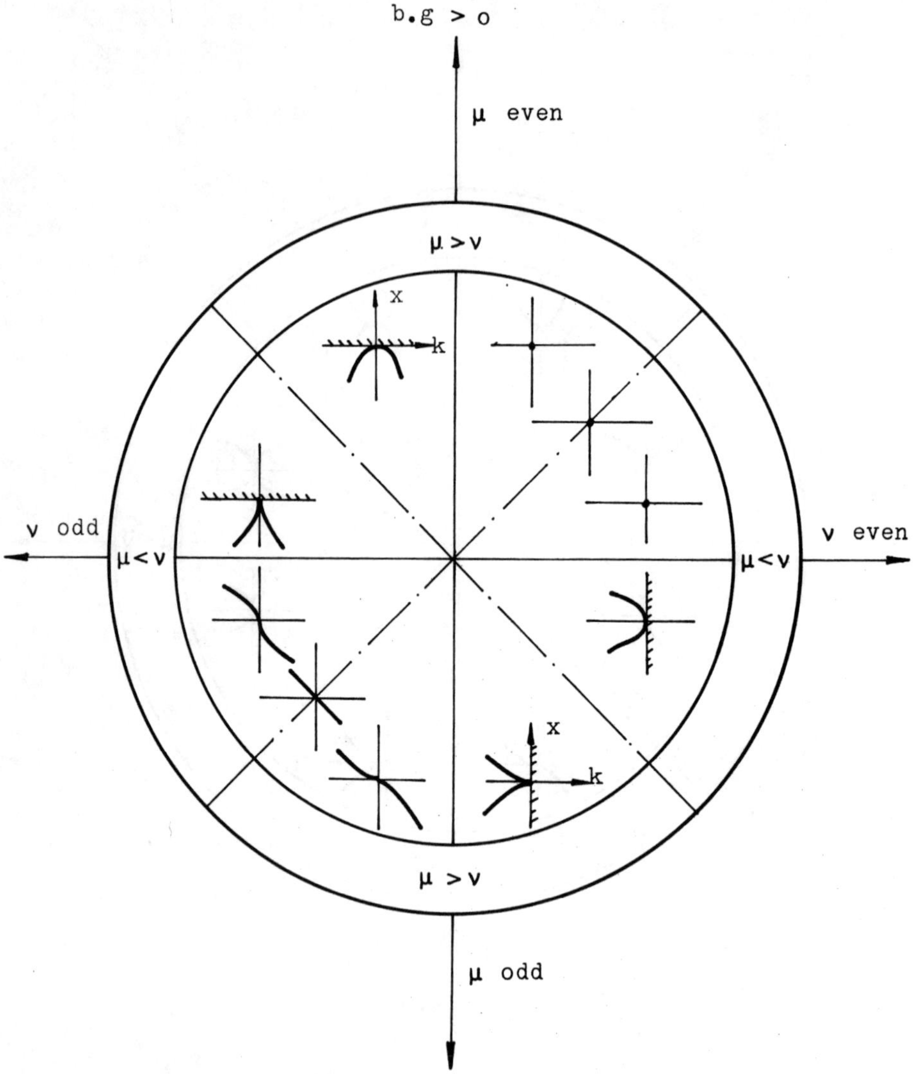

Figure 2.5.b: Elementary real branching diagrams of (2.56).

References

[2.1] D.T.Whiteside, The Mathematical Papers of Isaac Newton, Cambridge University Press, Vol.III, 1969.

[2.2] P.G.Aizengandler, On the branching theory of small solutions of nonlinear equations, Uchen. Zap. Moscov. Obl. Ped. Inst., 166,No.1o, 1966.

[2.3] M.M.Vainberg, V.A.Trenogin, Theory of branching of solutions of nonlinear equations, Noordhoff 1974.

[2.4] M.A.Krasnoselskii et al.,Approximate solutions of operator equations, Noordhoff, 1972.

[2.5] H.Hahn Existence Criteria for Bifurcations as Stability Criteria for Critical Nonlinear Systems, Springer Lecture Notes in Biomath., 1978.

[2.6] H.Hahn, Zur Theorie und Technik singulärer Regelkreise, Habilitationsschrift, Chapters III and XIV, Universität Tübingen, Fachbereich Physik, 1977/78.

III. The Exponent Diagram

1. Statement of the problem

In chapter II, the technique of the Newton Diagram for constructing all "small solutions" $x(k)$ of the equation

$$0 = F(x,k) = \sum_{i=0}^{q} \sum_{j=0}^{n_i} a_j^i . x^j . k^{r_i} \quad , \qquad (3.1)$$

where

$$i,j,n_i \in \mathbb{N} \;;\; r_i \in \mathbb{N} \quad \text{or} \quad r_i \in \mathbb{Q} \;;\; a_j^i, k \in \mathbb{R}^1 \text{ and } x \in \mathbb{C}^1,$$

has been illustrated. These interpretations were based on the formal theorems presented in the appendix A.1 and A.2 .

In practice very often not only the "small" solutions, but also other solutions of the equation (3.1), for instance its "large" solutions or even all of its solutions, are of some interest.

Within this chapter, the technique of the Newton Diagram will be extended towards the technique of the "Exponent Diagram" which not only allows to analyse and to construct all small solutions of (3.1) but also all of its large solutions.
In this context large solutions of equation (3.1) are defined by the relation

$$\lim_{k \to \infty} x(k) = \infty \quad , \qquad (3.2)$$

whereas small solutions of (3.1) satisfy the relation

$$\lim_{k \to 0} x(k) = 0 \quad . \qquad (3.3)$$

In order to construct all solutions of (3.1), it is useful to rewrite this equation in different forms where two assumptions are made :

(i) Equation (3.1) may be represented as a finite series, and
(ii) $r_i \in \mathbb{N}$.

Following the same lines as in the appendix A.2, the second restriction may easily be overcome. On the other hand, the assumption (i) is necessary for constructing all solutions of equation (3.1).

Then we have

$$F(x,k) = \sum_{i=0}^{q} \sum_{j=0}^{n_i} k^{r_i} . a_j^i . x^j = \sum_{j=0}^{n} \sum_{i=0}^{q_j} x^j . a_j^i . k^{r_i} \tag{3.4}$$

$$= a. \prod_{j=1}^{n} (x - x_j(k)) \qquad ,$$

where

$$r_i, i, j, n_i, q_j \in \mathbb{N} \quad ; \quad 0 = r_0 < r_1 < r_2 < \ldots < r_q$$

and

$$q := \sup_j q_j \quad , \quad n := \sup_i n_i \quad ; \quad k, a_j^i \in \mathbb{R}^1$$

or the two equivalent representations

$$F(x,k) = \sum_{i=0}^{q} k^{r_i} . Q_i(x) \qquad , \tag{3.5a}$$

where

$$Q_i(x) := \sum_{j=0}^{n_i} a_j^i . x^j = a_i . \prod_{j=1}^{n_i} (x - x_j^i) \ , \ i=0,1, \ \ldots \ ,q \tag{3.5b}$$

and

$$F(x,k) = \sum_{j=0}^{n} x^j . R_j(k) \qquad , \tag{3.6a}$$

where

$$R_j(k) := \sum_{i=0}^{q_j} k^{r_i} . a_j^i = a_j . \prod_{i=1}^{q_j} (k - k_j^i), \ j=0,1, \ \ldots \ ,n \, . \tag{3.6b}$$

2. Construction of the Exponent Diagram

In analogy to the deliberations concerning the Newton Diagram, the Exponent Diagram corresponding to (3.4) may be constructed by applying the following steps (compare Fig. 3.1a and 3.1b):

(i) Let $e(x) := j$ and $e(k) := i := r_i$ be the exponents of the variables x and k of the various terms of equation (3.4). Then a coordinate system is to be drawn with j as abscissa and i as ordinate.

(ii) All points (j,i) corresponding to the terms $x^j . k^i$ of equation (3.4) are to be inserted into the coordinate system e(x)-e(k).

(iii) A convex polygon of minimal length is to be drawn by connecting some of these points such that all points are encircled by the polygon. This polygon will be called EXPONENT POLYGON in what follows.

(iv) The Exponent Polygon is divided into eight segments termed I to VIII each of which consists of straight lines with slopes $\alpha_{I.1}$, $\alpha_{I.2}$, ..., $\alpha_{VIII.1}$, $\alpha_{VIII.2}$, ..., where $\beta_j := \tan\alpha_j$ (comp. Fig. 3.1b). The final figure will be called EXPONENT DIAGRAM .

(v) Let l_o be the degree of degeneration of (3.4) (comp. Definition A.1 of the appendix A).
Let e_I to e_{VIII} or $\bar{e}_I$ to $\bar{e}_{VIII}$ be the lenght of the projection of the segments I to VIII onto the abscissa e(x) and ordinate e(k) respectively. The constants l_o and e_I to $\bar{e}_{VIII}$ have to be inserted into the Exponent Diagram.

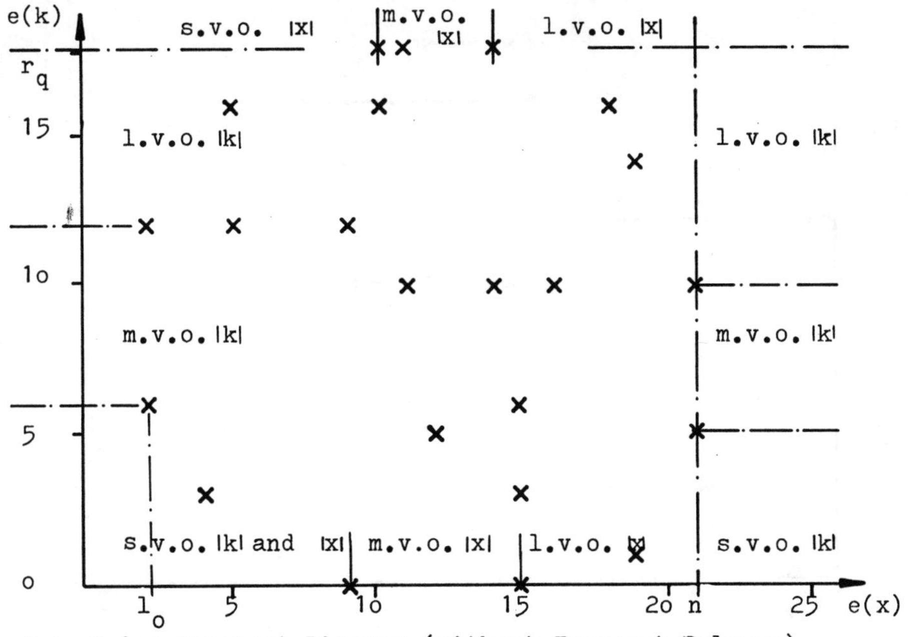

Fig. 3.1a: Exponent Diagram (without Exponent Polygon)

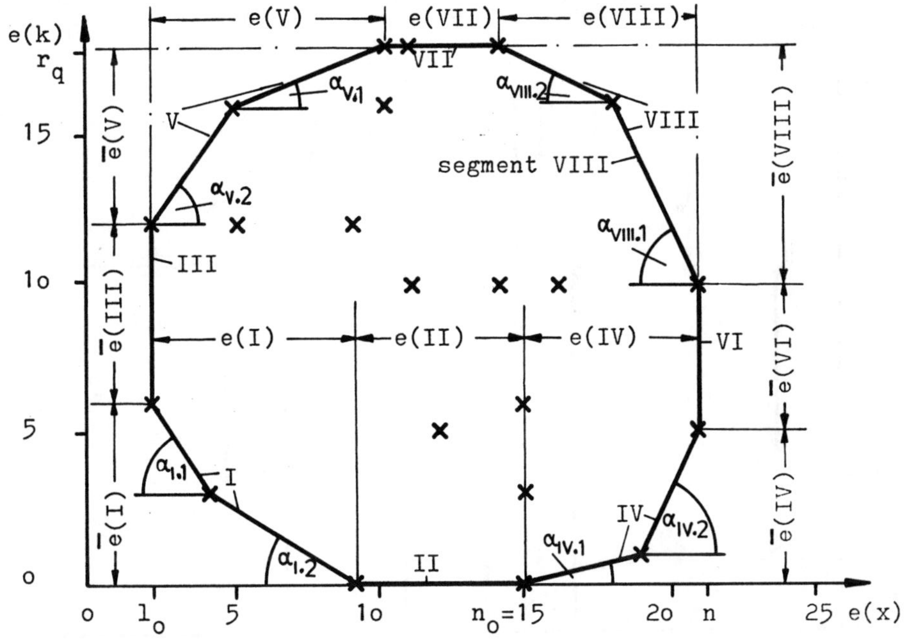

Fig.3.1b: Exponent Diagram and Exponent Polygon (s.v.o. = small values of ,l.v.o. = large values of, m.v.o. = mediumsized values of).

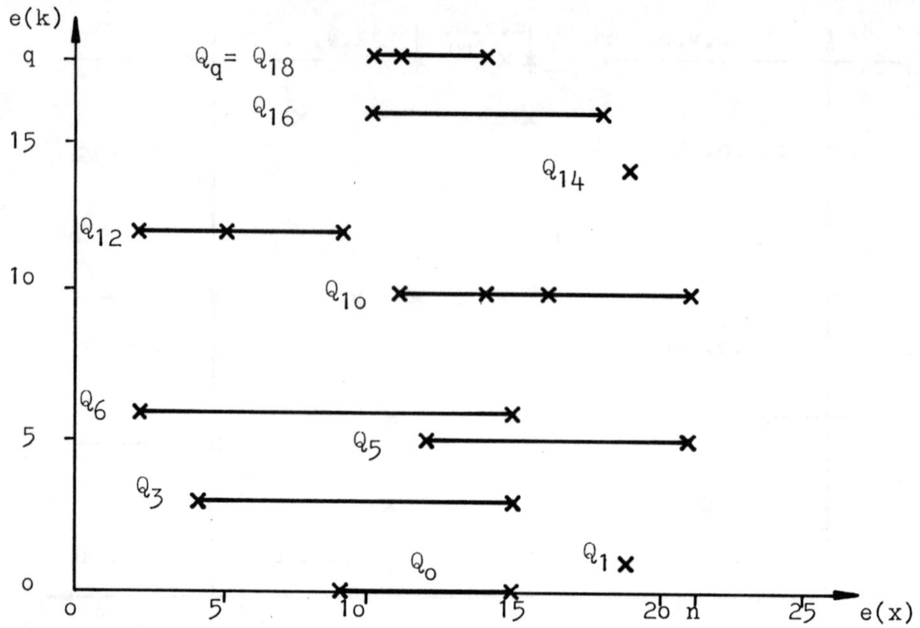

Fig.3.2a: Exponent Diagramm corresponding to equation (3.5)

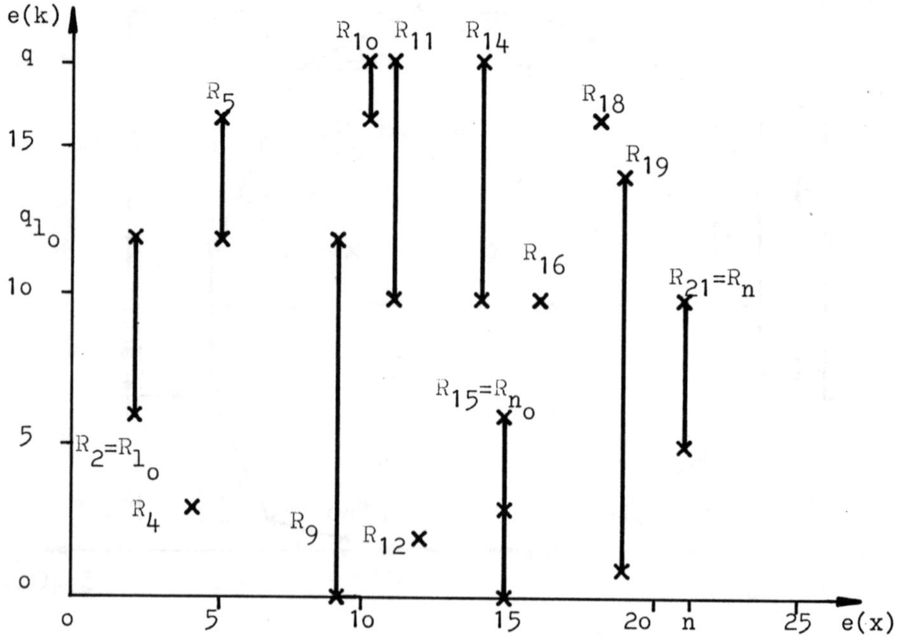

Fig.3.2b: Exponent Diagram corresponding to equation (3.6)

Then the following relations are obvious :

$$e_{III} = e_{VI} = 0 \quad ; \quad \bar{e}_{II} = \bar{e}_{VII} = 0 \tag{3.7a}$$

and

$$\alpha_{III} = \alpha_{VI} = \pi/2 \quad ; \quad \alpha_{II} = \alpha_{VII} = 0 \ . \tag{3.7b}$$

A comparison between chapter II and chapter III shows that the segment I of the Exponent Polygon is identical to the Newton Polygon corresponding to (3.4) .

3. Interpretation of the segments I to VIII of the Exponent Polygon (Fig.3.1b)

As will be shown now, each of the various segments I to VIII of the Exponent Polygon characterizes a class of solutions x(k) resp. k(x) of the equation (3.4).

Segment I :

According to chapter II, the segment I of the Exponent Polygon is identical to the Newton Polygon of (3.4). Therefore it describes all the small solutions of (3.4) which for small values of $|k|$ may be written in the form

$$x_j(k) = \gamma_j . k^{\beta_j} + o(|k|^{\beta_j}) \quad , \quad \beta_j = \tan\alpha_{I.j} \tag{3.8a}$$

, where $\beta_j > 0$ for $j = 1, \dots , e_I$ (some of which may be equal), $\beta_j = 0$ for $j = e_I + 1, \dots , e_I + l_o$ and γ_j is a non vanishing root of the corresponding supporting polynomial (compare chapter II).

Segment II :

The polynomial $Q_o(x)$ has exactly $l_o + e_I$ vanishing roots and e_{II} non vanishing roots called x_j^o $(j = 1, \ldots, e_{II})$, some of which may appear as complex conjugate pairs, and some of which may be equal. Using the transformation

$$x \longrightarrow x - x_j^o =: x' \tag{3.9b}$$

, the segment II of the Exponent Polygon of the function F(x,k) is transferred into the segment II of the Exponent Polygon of the function

$$_jF_{II}(x',k) := F(x - x_j^o\,;\,k) \quad . \tag{3.10b}$$

The corresponding solutions have the form

$$x_j(k) = x_j^o + \gamma_j . k^{\tilde{\beta}_j} + o(|k|^{\tilde{\beta}_j}) \, , \quad \tan \tilde{\beta}_j = \tilde{\alpha}_{I,j} \tag{3.8b}$$

, $\tilde{\beta}_j > 0$ for $j = 1, \ldots, e_{II}$ some of which may be equal

, where $\tilde{\alpha}_{I.j}$ is taken from the segment I of the Exponent Polygon of the function $F(x',k)$, and γ_j are the non vanishing solutions of the corresponding supporting polynomial (compare chapter II).
Clearly, the solutions (3.8b) are neither small nor large solutions..They are called mediumsized solutions.

Segment III :

The segment III describes solutions of equation (3.4) of the form

$$x_j(k) = \gamma_j . k'^{\tilde{\beta}_j} + o(|k'|^{\tilde{\beta}_j}) \, , \quad \tilde{\beta}_j = \tan \tilde{\alpha}_{I.j} \tag{3.8c}$$

, where k' is defined by the transformation

$$k \longrightarrow k - k_j^o =: k' \tag{3.9c}$$

and k_j^o is one of the at most $\bar{e}_{III}$ non vanishing real solutions of the expression $R_{1_o}(k)$ (compare Fig. 3.2a).
The coefficients γ_j are roots of the corresponding supporting polynomial. The segment III of $F(x,k)$ is transformed by the relation (3.9c) into the segment I of the function

$$ {}_jF_{III}(x,k') := F(x,k - k_j^o) \quad . \tag{3.10c}$$

Segment IV :

The segment IV of the Exponent Diagram of (3.4) is connected to the solutions satisfying the relation

$$ \lim_{k\to 0} x(k) = \infty \quad . \tag{3.11a}$$

These solutions rise at infinity for $k = 0$ or they tend to infinity for decreasing $|k|$. There exist exactly $e_{IV} = n - n_o$ (complex) solutions of this type.
Using the transformation

$$ x \longrightarrow 1/x =: x' \tag{3.9d}$$

, the segment IV of $F(x,k)$ is transferred into the segment I corresponding to the function

$$ F_{IV}(x',k) := \sum_{j=0}^{n} \sum_{i=0}^{q_j} a_j^i . x'^{j'} . k^i \ , \ j' := n - j \ . \tag{3.10d}$$

Let ${}_{IV}a_j^i$ be the coefficients of $F(x,k)$ corresponding to points on the segment IV of (3.4), where $i \in I_{IV}$. Then taking into consideration only these ponts, we have

$$ \tilde{F}(x,k) = x^n . \sum_{j=n_o}^{n} x'^{j'} . \sum_{i \in I_{IV}} {}_{IV}a_j^i . k^{r_i} = x^n . \tilde{F}_{IV}(x',k) \tag{3.12a}$$

, where

$$\tilde{F}_{IV}(x',k) := \sum_{j'=0}^{n-n_o} x'^{j'} \cdot \sum_{i\in I_{IV}} {}_{IV}a^{i}_{n-j'} \cdot k^{r_i}$$

$, j' := n - j \ , \quad j' = 0 \ \text{ for } \ j = n \ \text{ and } \ j' = n-n_o \ \text{ for } \ j = n_o .$

As a result, the corresponding solutions of (3.4) take for small values of $|k|$ the form

$$x_j(k) = \gamma_j^{-1} \cdot k^{-|\beta_j|} + o(k^{-|\beta_j|}) \ , \ \beta_j = \tan \alpha_{IV.j} \tag{3.8d}$$

, where the coefficients γ_j are roots of the supporting polynomial corresponding to (12a). Due to the transformation (3.9d), the segment IV of $F(x,k)$ and the segment I of (3.1od) change the roles (compare Fig. 3.3a).

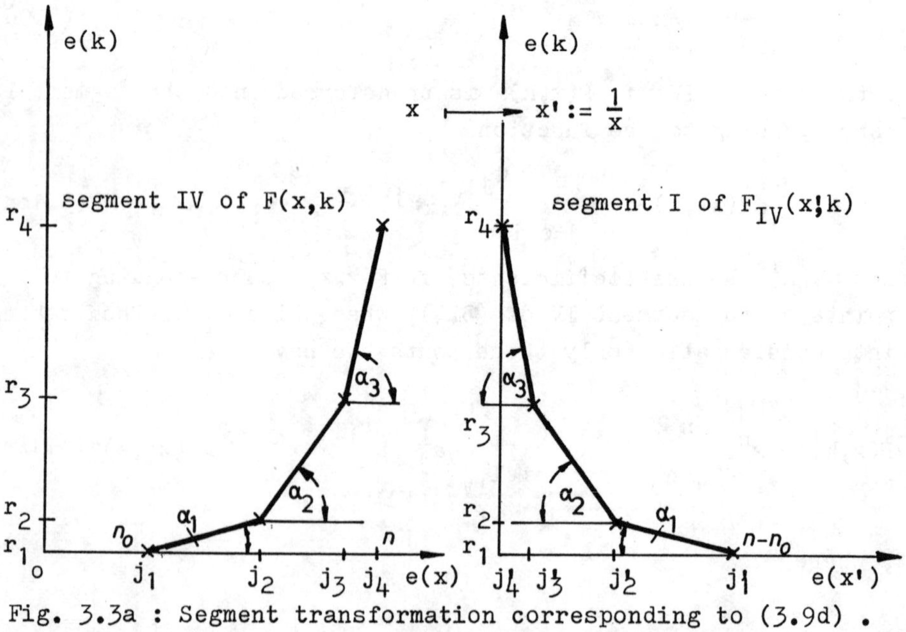

Fig. 3.3a : Segment transformation corresponding to (3.9d) .

Segment V :

Applying the transformation

$$k \longrightarrow 1/k =: k' \tag{3.9e}$$

to $F(x,k)$, the segment V of the Exponent Diagram is transferred into the segment I of the function $F_V(x,k')$,

$$F_V(x,k') := \sum_{i=0}^{q} \sum_{j=0}^{n_i} a^i_j . x^j . k'^{i'} , \quad i' := q - i \tag{3.10e}$$

(compare Fig. 3.3b). Let ${}_V a^i_j$ be the coefficients of $F(x,k)$ corresponding to the points on the segment V of (3.4) and let $j \in J_V$. Then taking into account only those terms of (3.4) which correspond to points on the segment V, we have

$$\tilde{F}(x,k) = k^q . \sum_{i=q_{1_o}}^{q} k'^{i'} . \sum_{j \in J_V} . {}_V a^i_j . x^j = k^q . \tilde{F}_V(x,k')$$

resp. (3.12b)

$$\tilde{F}_V(x,k') := \sum_{i'=0}^{q-q_{1_o}} k'^{i'} . \sum_{j \in J_V} {}_V a^{q-i'}_j . x^j \quad .$$

, where

$$i' := q - i \quad \text{and} \quad i' = 0 \quad \text{for} \quad i = q \quad .$$

As a result, the corresponding solutions of (3.4) take for small values of $|k'|$ the form

$$x_j(k) = \gamma_j . k'^{\beta_j} \quad ; \quad \beta_j := \tan \alpha_{V.j} \quad . \tag{3.8e}$$

The coefficients γ_j are the non vanishing roots of the supporting polynomials corresponding to (3.12b). There exist exactly e_V of these solutions. By applying the transformation (3.9e) to (3.4), the segment V of (3.4) and the segment I of (3.12b) change the roles.

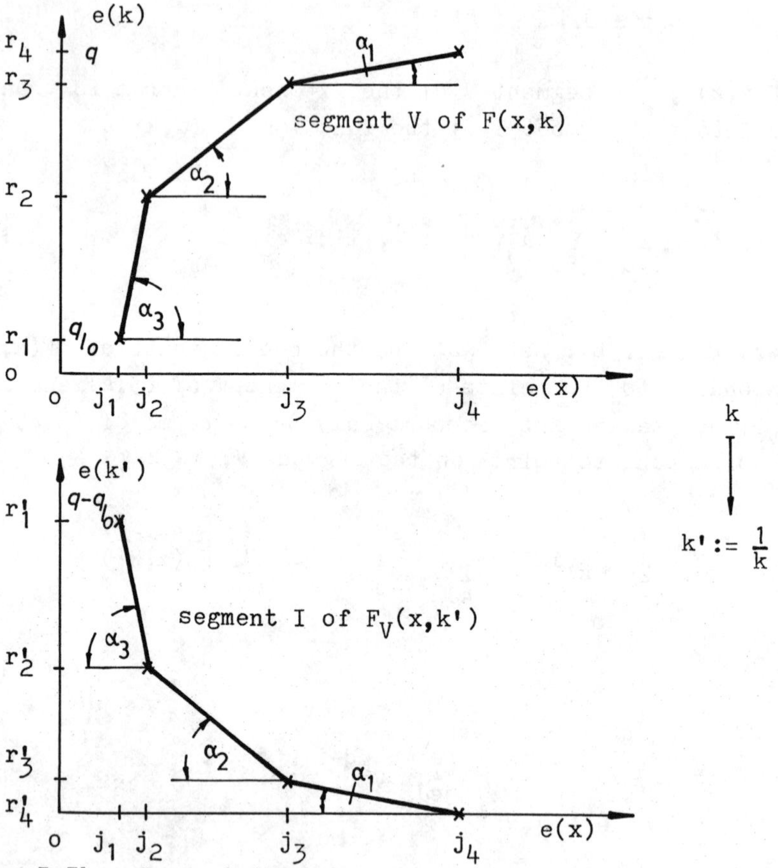

Fig. 3.3b : Segment transformation corresponding to (3.9e)

<u>Segment VI :</u>

Using the transformation

$$x \longrightarrow 1/x =: x' \quad , \quad k \longrightarrow k - k_n^j =: k' \quad , \quad j = 1,2, \ldots, \bar{e}_{VI} \qquad (3.9f)$$

, where k_n^j are the at most $\bar{e}_{VI}$ real roots of $R_n(k)$, the segment VI of the Exponent Diagram of $F(x,k)$ is transformed to the segment I of the function

$$ {}_jF_{VI}(x',k') := \sum_{j=0}^{n} \sum_{i=0}^{q_j} a_j^i . x'^{j'} . (k-k_n^1)^i, \; j':=n-j. \qquad (3.1of) $$

For small values of $|k'|$ the corresponding solutions take the form

$$ x_j(k) = \gamma_j^{-1} . k'^{-\beta_j} \quad , \quad \beta_j := \tan \alpha_{VI.j} \qquad (3.8f) $$

,where γ_j is one of the $\bar{e}_{VI}$ non vanishing roots of the supporting polynomial corresponding to the segment I of(3.1of).

Segment VII :

The segment VII of the Exponent Diagram corresponding to (3.4) may be transferred to the segment I of the function

$$ F_{VIII}(x',k') := \sum_{i=0}^{q} \sum_{j=0}^{n_i} a_j^i . (x-x_1^q)^j . k'^{i'} ; i' := q-i \qquad (3.1og) $$

by means of the transformation

$$ x \longrightarrow x - x_j^q =: x' \quad \text{and} \quad k \longrightarrow 1/k =: k' \qquad (3.9g) $$

, where

x_j^q is one of the e_{VII} **non** vanishing roots of $Q^q(s)$.
For large values of $|k|$ the corresponding e_{VII} solutions have the form

$$ x_j(k) = x_j^q + \gamma_j . k'^{\beta_j} \quad , \quad \beta_j = \tan\alpha_{VII.j} \qquad (3.8g) $$

, where γ_j are non vanishing roots of the supporting polynomial corresponding to (3.1og), and $\beta_j = \beta_{VII.j}$ is taken from the corresponding segment of the Exponent Diagram.

Segment VIII :

Applying the transformation

$$x \longrightarrow 1/x =: x' \ , \ k \longrightarrow 1/k =: k' \qquad (3.9h)$$

, the segment VIII of the function $F(x,k)$ is transferred into the segment I of the function

$$F_{VIII}(x',k') := \sum_{j=0}^{n} \sum_{i=0}^{q_j} a_{n-j'}^{q-i'} . x'^{j'} . k'^{i'} \qquad (3.10h)$$

(compare Fig. 3.3c). Let ${}_{VIII}a_j^i$ be the coefficients of $F(x,k)$ corresponding to the points on the segment VIII of (3.4), and let $j\epsilon J_{VIII}$ and $i\epsilon I_{VIII}$, where J_{VIII}, I_{VIII} are the corresponding index sets.
Then taking into account only those terms of (3.4) which correspond to points on the segment VIII we have

$$\tilde{F}(x,k) = x^n.k^q. \sum_{j\epsilon J_{VIII}} x'^{j'}. \sum_{i\epsilon I_{VIII}} a_j^i .k'^{i'} = x^n.k^q.\tilde{F}(x',k')$$

, where (3.12h)

$$\tilde{F}(x',k') = \sum_{j\epsilon J_{VIII}} x'^{j'}. \sum_{i\epsilon I_{VIII}} a_{n-j'}^{q-i'}.k'^{i'} \quad \text{and}$$

$$i' := q - i \quad , \quad i' = 0 \quad \text{for} \quad i = q \quad .$$

For large values of $|k|$, the corresponding solutions of (3.4) are of the form

$$x_j(k) = \gamma_j.k^{\beta_j} \qquad (3.8h)$$

, where $\beta_j := \tan\alpha_{VIII.j}$ and γ_j^{-1} is a non vanishing root of the corresponding supporting polynomial .

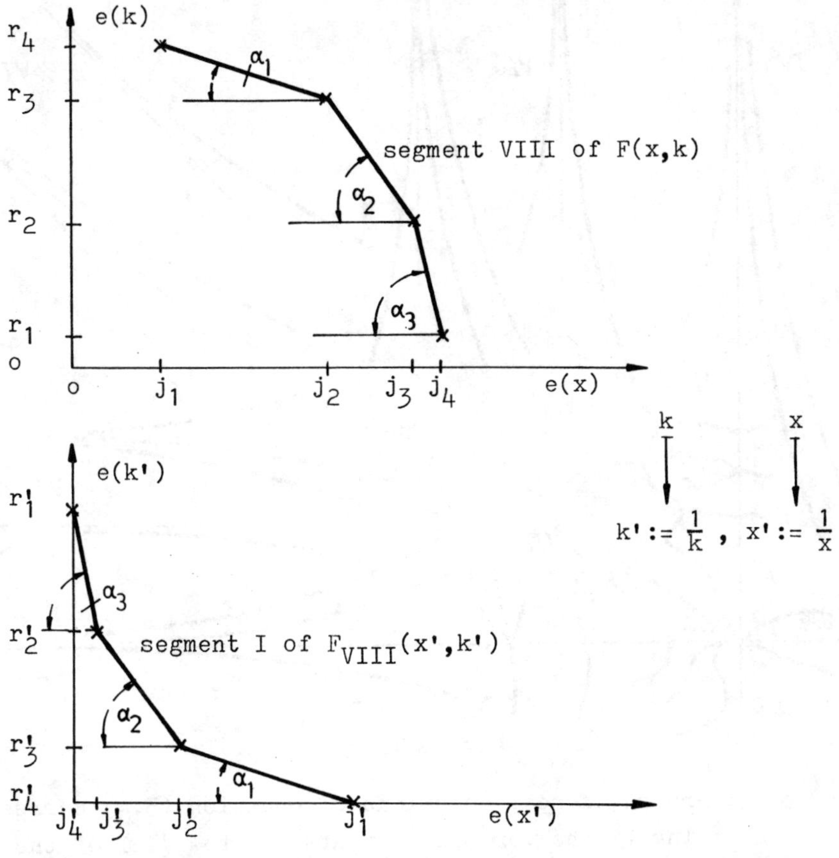

Fig. 3.3c : Segment transformation corresponding to (3.9h)

The <u>real</u> solutions of equation (3.4) corresponding to the segments I to VIII of the Exponent Diagram are sketched in figure 3.4 .

In case the function F(x,k) is a pseudopolynomial instead of a polynomial (comp. appendix A.2), some of the segments II to VII may be of infinite length.

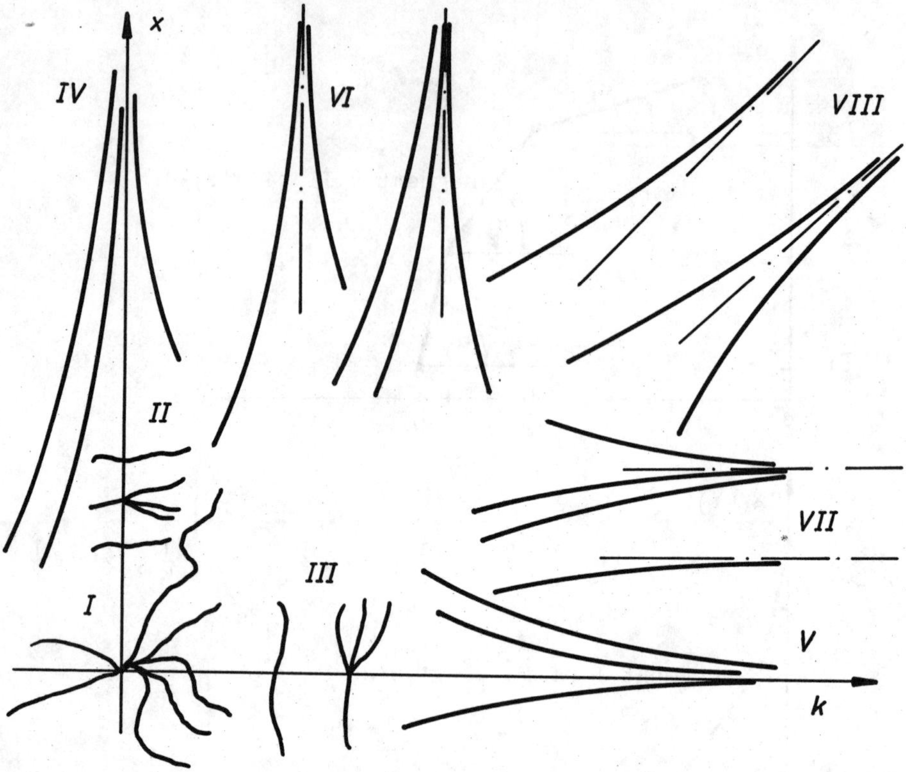

Fig. 3.4 : Groups of real solutions of equation (3.4) corresponding to the various segments I to VIII of the Exponent Diagram .

IV. Higher Order Root Locus Construction Rules

1. Statement of the problem

The classical root locus method developed by Evans [4.1] is a well approved graphical means of determining the poles of the transfer function of a single input closed-loop control system if the poles and zeros of the corresponding open-loop transfer function are known.
The root locus is a picture of the behaviour of the closed loop poles as one of the (scalar) system parameters, usually the gain k, is varied.

Formally this technique is based on the characteristic polynomial of the closed-loop system

$$P(s,k) = \sum_{j=0}^{n} a_j . s^j + k . \sum_{i=0}^{m} b_i . s^i \,, \tag{4.1}$$

where $s \in \mathbb{C}^1$; a_j , b_i , $k \in \mathbb{R}^1$ and $n,m \in \mathbb{N}$.

This polynomial is <u>linear</u> in the parameter k . Therefore this technique will be called "<u>first order</u> root locus technique" in what follows.

In connection with complex practical control problems there exists a variety of questions which are based on a characteristic polynomial of the related system of the following form

$$P(s,k) = \sum_{i=0}^{q} \sum_{j=0}^{n_i} a_i^j . s^j . k^{r_i} \,, \tag{4.2}$$

where i , $j \in \mathbb{N}$, $r_i \in \mathbb{N}$ or $r_i \in \mathbb{Q}$ and $a_i^j \in \mathbb{R}^1$.

Following the same reasoning as in chapter III, we will restrict our investigations, without loss of generality, to polynomials ($r_i \in \mathbb{N}$). Contrary to equation (4.1), the equation (4.2) is <u>non linear</u> in the parameter k . Therefore, the corresponding root locus method will be called "<u>higher order</u> root locus technique"in what follows.

Characteristic polynomials of the form of equation (4.2) may occur in connection with linear multiloop control systems as well as in non linear control systems. In the latter case, in general, we have $r_i \in Q^+$. In both cases the appearance of a non linear parameter dependency may have quite different reasons.

Linear multiloop control sytems with a characteristic polynomial of the form (4.2) may, among others, arise from situations, where

(i) subsystems of a control system offer strong symmetrical couplings [4.2, 4.3],

(ii) the synthesis of an optimal control system requires a compromise between the reaction velocity of the system and the energy required by the controller [4.4],

(iii) the parameter sensitivity of the control system or network is of basic importance,

(iv) the reliability and the robustness of the system in the presence of simultaneous failures in some sensors or controllers is of some interest,

(v) the starting-, repair-, or maintenance procedure of the control system is performed by changing various system parameters simultaneously and proportional to each other.

System equations of the form (4.2) may arise in connection with non linear control systems in situations, where

(i) a non linear control system incorporating one or several non linear elements is treated by means of the "describing root locus technique"[4.5],

(ii) the stationary branching solutions of a non linear control system are of some interest [4.6], and

(iii) the periodic branching solutions of a non linear control system are of some importance [4.7].

These examples announce already the need for systematic construction rules of higher order root locus plots corresponding to equation (4.2).

Moreover, because of the non linear parameter dependence of equation (4.2), the relations between the prescribed specifications in system design and the control parameters are far from being transparent in view of the variety of geometric phenomena of the corresponding root locus plots.

Furthermore these various phenomena may make it hard to decide without strong root locus construction rules, whether the numerical results in connection with the eigenvalue computation of large scale systems are correct or not.

Summarizing, we may state that higher order root locus construction rules are needed,

(i) to superwise the numerical eigenvalue computations and the corresponding parameter dependence of large scale systems,

(ii) to produce insight into the internal structure of complex control systems, and

(iii) to deliver rules and concrete hints in connection with a systematic design of multiloop control systems.

Various investigations are known from literature which tend to derive higher order root locus construction rules in analogy to Evan's approach [4.3 , 4.8 , 4.9 , 4.1o , 4.11 , 4.12] .

Some of the rules derived within these investigations are not correct [4.3], others are incomplete [4.8 , 4.9 , 4.1o , 4.11] or too complicated to be applicable in practice.

Using the technique of the <u>Exponent Diagram</u> derived in chapter III ,the higher order root locus construction rules will be derived systematically in what follows.

On the basis of this procedure, the variety of different <u>algebraic cases</u> in connection with the solutions of equation (4.2) is transformed into a simple <u>geometric pattern.</u> Therefore this procedure has a strong appeal to the intuitive and graphic way of ingeneering thinking. Moreover it proves to be a natural generalization of the classical (first order) root locus technique. In view of the increasing complexity of equation (4.2) compared to equation (4.1), the higher order root locus construction rules must be expected to be complexer than the classical ones.

2. Derivation of the higher order root locus construction rules.

In analogy to chapter III, it is useful to start the investigation of the solutions of equation (4.2) on the basis of the following equivalent representations of this equation :

$$P(s,k) = \sum_{i=0}^{q} \sum_{j=0}^{n_i} k^{r_i} . a_j^i . s^j = \sum_{j=0}^{n} \sum_{i=0}^{q_j} s^j . a_j^i . k^{r_i}$$

$$= \prod_{j=1}^{n} (s - s_j(k)) \quad , \qquad (4.3)$$

where

$$i, j, n_i, q_j \in \mathbb{N}; r_i \in \mathbb{N} \text{ or } r_i \in \mathbb{Q}; 0 = r_0 < r_1 < \dots < r_q \quad \text{and}$$

$$q := \sup_j q_j \quad , \quad n := \sup_i n_i \quad , \quad s \in \mathbb{C} \; ; \; k \, , \, a_j^i \in \mathbb{R}^1 \qquad (4.4)$$

, or

$$P(s,k) = \sum_{i=0}^{q} k^{r_i} . Q_i(s) \qquad (4.5)$$

, where

$$Q_i(s) = \sum_{j=0}^{n_i} a_j^i . s^j = a_i \cdot \prod_{j=1}^{n_i} (s - s_j^i) \quad ; \quad i \in \{0, \dots, q\}$$

and

$$P(s,k) = \sum_{j=0}^{n} s^j . R_j(k) \qquad (4.6)$$

, where

$$R_j(k) := \sum_{i=0}^{q_j} k^{r_i} . a_j^i = a_j . \prod_{i=1}^{q_j} (k - k_j^i) ; \; j \in \{0, \dots, n\} \quad .$$

On the basis of the investigations of chapter III, the root locus construction rules corresponding to (4.3), (4.4), (4.5) and (4.6) will be derived step by step. The factor k will be interpreted as a feedback factor of the controlsystem in what follows.

STEP 1 : Construction of the Exponent Diagram and of the Exponent Polygon corresponding to equation (4.3).

This step has to be performed in analogy to chapter III.2 by replacing the variable x by means of the variable s (compare Fig. 3.1a and Fig. 3.1b).

Note 1 :

The Exponent Polygon corresponding to equation (4.1) may take any one of the typical forms sketched in figure 4.1. It contains at most four of the eight segments of figure 3.1b at the same time, and each of these segments consists of exactly one straight line.

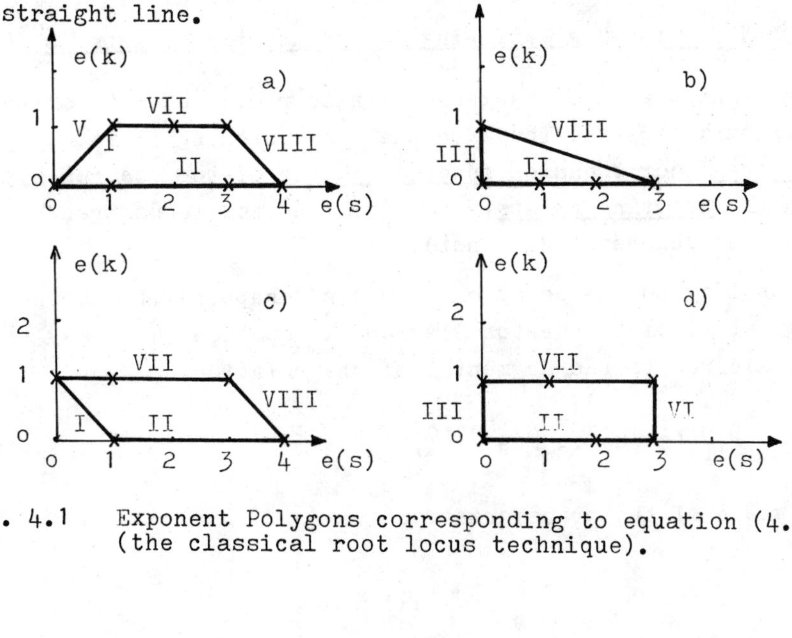

Fig. 4.1 Exponent Polygons corresponding to equation (4.1) (the classical root locus technique).

STEP 2 : Interpretation of the various segments of the Exponent Polygon with respect to the root locus concept (compare Fig. 3.1b).

Referring to chapter III.3 , the segments I to VIII will be attached to the different notions (branch, asymptote, zero, pole, ...) of the root locus concept.

SEGMENT I : The starting point $s = o$ $(k = o)$

There exist $(l_o + e_I)$ vanishing roots s_j^o of the polynomial $Q_o(s)$ of equation (4.5) which are the vanishing poles of the open loop transfer function matrix of the control system. Therefore, the point $s = o$ is a starting point of exactly e_I of the root locus branches for $k = o$ which will leave these points for values $|k| \neq o$. Exactly l_o of the root locus branches will be fixed at this point as degenerated branches for all real values of k .

Note 2 :

The vanishing roots of $Q_o(s)$ are emphasized here because all roots of the various polynomials Q_j and R_i may be reduced to vanishing zeros of some transformed polynomials $\tilde{Q}_j$ and $\tilde{R}_i$ in the following investigations.

SEGMENT II : Non vanishing finite starting points s_j^o $(k=o)$.

There exist e_{II} non vanishing finite roots $s_j^o \neq o$ of the polynomial $Q_o(s)$. They are the non vanishing finite poles of the open loop transfer function matrix of the control system and the starting points of e_{II} of the root locus branches (whether degenerated or not).

In analogy to the relation (3.9b) of chapter III.3 , the segment II of the Newton Diagram of equation (4.5) may be transferred to the segment I of the relation

$$P_{II}(s',k) := P(s - s_j^o, k) \tag{4.7a}$$

by means of the transformation

$$s \longrightarrow s - s_j^o = : s' \tag{4.8a}$$

Note 3 :

As will be shown in connection with the examples of the following chapters which illustrate the higher order root locus technique, the transformations (4.8) have not to be performed in explicit form, in general. The amound of work in connection with this step depends on the fact whether s_j^o is a simultaneous zero of $Q_o(s)$ and $Q_1(s)$ and of the various polynomials $Q_j(s)$ or not, where $j = o, 1, 2, \dots, q$.

SEGMENT III : Branches which run through the point $s = o$ for finite non vanishing real values $k_{1_o}^j$ of k.

There exist exactly $\bar{e}_{III}$ finite non vanishing roots $k_{1_o}^j$ of the polynomial $R_{1_o}(k)$ some of which, e.g. $\tilde{e}_{III}$ of which may be real, where $\tilde{e}_{III} \leq \bar{e}_{III}$. Then the point $s = o$ will be crossed $\tilde{e}_{III}$ times by root locus branches for $k_{1_o}^j \neq o$, $k_{1_o}^j$ real, where each branch may reach this point several times. Again, the segment III of (4.6) may be transferred to the segment I of the relation

$$P_{III}(s,k') : = P(s,k - k_{1_o}^j) \tag{4.7b}$$

by means of the transformations

$$k \longrightarrow k - k_{1_o}^j = : k' \tag{4.8b}$$

, and the statement of the note 3 is valid for Q_o and Q_j replaced by R_{1_o} and R_j .

SEGMENT IV : Infinite starting points (k = o).

There exist exactly e_{IV} root locus branches which start at infinity ($s = \infty$) for $k = o$. The segment IV of (4.5) is transferred to the segment I of the relation

$$P_{IV}(s',k) := P(s,k).s^{-n} \qquad (4.7c)$$

by means the transformation

$$s \longrightarrow s' := 1/s \ . \qquad (4.8c)$$

Note 4 :

Infinite starting points of root locus branches only occur in control systems if the derivatives of some of the state-variables of the first order differential equations are multiplied by the factor k [4.13, 4.14, 4.15]. In this case, the dimension of the system may change abruptly as the factor k takes specific real values k" of k. The relation $n_o < n$ (comp. Fig. 3.1b) is a necessary and sufficient condition for the occurrence of $(n - n_o)$ infinite starting points. Theoretical situations of this type are treated formally within the framework of the "Singular Perturbation Theory". From the control theoretical point of view, these systems contradict locally (for k = k") the well known realization conditions of linear systems.

SEGMENT V : Branches which meet the point s = o for $k = \infty$ (vanishing sinks of the system).

There exist e_V root locus branches which meet the point s = o for infinite values of k, where each branch is counted according to its multiplicity. The $(l_o + e_V)$ vanishing zeros s_j^q of the polynomial $Q_q(s)$ are the vanishing sinks of the transfer function matrix of the multiloop control system (compare note 5, [4.16] and chapter VII). The segment V of (4.5) may be transferred to the segment I of the relation

$$P_V(s,k') := P(s,k).k^q \qquad (4.7d)$$

by means of the transformation

$$k \longrightarrow k' := 1/k \ . \qquad (4.8d)$$

Again, the statement of the note 3 is valid.

Note 5 :

The sinks of the equation (4.5) are the finite end points, the finite goals of the root locus branches for $k = \infty$. They correspond to the zeros of a single loop control system.

SEGMENT VI : Branches which tend to infinity for finite real values of k.

There exist $\bar{e}_{VI}$ non vanishing finite roots k_n^j of $R_n(k)$, $\tilde{e}_{VI}$ of which are real, where $\tilde{e}_{VI} \leq \bar{e}_{VI}$. The "infinite point" $s = \infty$ is reached exactly $\tilde{e}_{VI}$ times by a root locus branch for finite real values k_n^j of k. The segment VI of (4.6) is transferred into the segment I of the relation

$$P_{VI}(s',k') := P_j(s,k - k_n^j).s^{-n} \qquad (4.7e)$$

by means of the transformation

$$s \longrightarrow s' := 1/s \quad \text{and} \quad k \longrightarrow k' := k - k_n^j . \qquad (4.8e)$$

Note 6 :

The occurence of the segment VI in a control system indicates that there may exist root locus branches which tend to infinity for finite real values of k (compare note 4).

SEGMENT VII : Finite non vanishing sinks $s_j^q = 0$ of the system (for $k = \infty$).

There exist e_{VII} non vanishing roots s_j^q of the polynomial $Q_q(s)$ which may be simple or multiple, real or complex. These roots are the finite non vanishing sinks, the goals of e_{VII} of the root locus branches, some of which may be degenerated or multiple (compare note 5).
The segment VII of (4.5) is transferred to the segment I of the relation

$$P_{VII}(s',k') := P(s - s_j^q ,k).k^q \qquad (4.7f)$$

by means of the transformation

$$s \longrightarrow s' := s - s_j^q \quad \text{and} \quad k \longrightarrow k' := 1/k \,. \tag{4.8f}$$

Again, the statement of the note 3 is valid.

SEGMENT VIII : Branches which tend to infiniy for $k = \infty$.

There exist e_{VIII} root locus branches which tend to infinity for $k = \infty$. This is easily seen by transferring the segment VIII of (4.3) into the segment I of the relation

$$P_{VIII}(s',k') := P(s,k).s^n.k^q \tag{4.7g}$$

by means of the transformation

$$s \longrightarrow 1/s \quad \text{and} \quad k \longrightarrow 1/k \tag{4.8g}$$

(compare chapter III. 3, Fig. 3.3c).

Note 7 :

The constant l_o corresponding to the polynomial $P(s,k)$ determines the number of the complete pole-sink-compensations at $s = o$ [4.16]. In this case, $s = o$ is a common root of order l_o of all of the polynomials Q_o to Q_q, and the point $s = o$ is a degenerated root locus branch of multiplicity l_o.
In analogy, the constant l_o corresponding to the transformed polynomial $P_j(s',k')$ of any of the equations (4.7) determines the number of the complete pole-sink-compensations near a point $s = s'$, where s' is defined by the corresponding equation (4.8). In this case, a term $(s-s')$ may be factored out from equation (4.3).
The proof of this statement follows directly from the fact that a complete pole-sink-compensation of the polynomial $P(s,k)$ is defined by the existence of a common zero s' of all of the polynomials $Q_o(s)$ to $Q_q(s)$.

In analogy to the preceeding comment, a common real zero k' of all of the polynomials R_{l_o} to R_n implies the existence of at least one trivial bifurcation $s(k)$ at $k = k'$. In this case, a term $(k - k')$ may be factored out from

equation (4.3).
From step 2 in connection with the considerations of chapter III.3 , the following results may be derived.

STEP 3 : Number of root locus branches (compare segment I, II and IV of Fig. 3.1b).

From Fig. 3.1b we have the relation

$$n = l_o + e_V + e_{VII} + e_{VIII} = l_o + e_I + e_{II} + e_{IV} \quad . \qquad (4.9)$$

Then there exist exactly n root locus branches, at least l_o of which are degenerated to an isolated point.

STEP 4 : Starting points of root locus branches for k = o (compare segment I, II, and IV of Fig. 3.1b).

Exactly $n_o = l_o + (e_I + e_{II})$ root locus branches start at roots s_j^o of the polynomial $Q_o(s)$ for $k = o$, $(l_o + e_I)$ of which are vanishing roots (segment I and II).

Exactly $e_{IV} = (n - n_o)$ branches start at infinity for $k = o$ (segment IV). Exactly l_o branches are degenerated to the point $s = o$.

STEP 5 : Sinks of root locus branches for $k = \infty$, k real (compare segment V, VII and VIII).

a) Exactly $(e_V + e_{VII})$ root locus branches tend for increasing $|k|$ towards the sinks of the system which are defined as roots s_j^q of the polynomial $Q_q(s)$ (segment V and VII).

b) Exactly e_V of these non degenerated branches end in the common sink $s = o$. Some of these $e_V + e_{VII}$ root locus branches may be degenerated to a fixed isolated point $s_j^q \neq o$.

c) Exactly $e_{VIII} = (n - n_q)$ root locus branches tend to infinity for $k = \infty$, k real. The corresponding goals will be called "infinite sinks" (segment VIII).

STEP 6 : Symmetry of locus.

Because of the assumptions $s \in \mathbb{C}^1$ and $k, a^i_j \in \mathbb{R}^1$ of equation (4.3), the root-locus diagram is always symmetric with respect to the real axis $\mathrm{Re}\{s\}$ of the complex plane. This is easily seen by substituting real constant values for k into the equation (4.3). The complex roots of (4.3) always appear as complex conjugate pairs.

STEP 7 : Number of branches which tend to zero for finite non vanishing real values $k^j_{1_o}$ of k. (segment III).

The polynomial $R_{1_o}(k)$ has exactly $\bar{e}_{III}$ non vanishing roots some of which ($\tilde{e}_{III}$) may be real. Then in agreement with chapter III.3 and with step 2 of this chapter, $\tilde{e}_{III}$ of these root locus branches will meet the point $s = o$ for $k=k^j_{1_o}$.

Note 8 :

An analogous statement holds for the number of branches which tend towards a non vanishing finite root s^o_j of the polynomial $Q_o(s)$ by computing the corresponding polynomial $R_{1_o}(k)$ and the corresponding number $\tilde{e}_{III}$ of the transformed polynomial $P(s - s^o_j, k) =: P_{II}(s',k)$.

STEP 8 : Number of branches which tend to infinity for finite non vanishing values k^j_n of k. (segment VI).

The polynomial $R_n(k)$ has $\bar{e}_{VI}$ non vanishing roots, $\tilde{e}_{VI}$ of which are real. Then exactly $\tilde{e}_{VI}$ root locus branches tend to infinity for finite non vanishing real values k^j_n of k, where $\tilde{e}_{VI} \leq \bar{e}_{VI}$.

The results of the following step will be used in the proof of various of the subsequent steps.

STEP 9 : The small solutions s(k) of the polynomial (4.3) (segment I).

According to the Appendix A.1, the small solutions of (4.2)

are defined by the relation

$$\lim_{k \to 0} s(k) = 0 . \qquad (4.10)$$

In agreement with equation (3.8a) of chapter III.3, each straight line of the segment I defines a slope α_{Ij} and a variable $\beta := \beta_{Ij} = \tan \alpha_{Ij}$ which together determine a small solution of (4.3) of the form

$$s = \gamma . k^{\beta}, \ \beta = \tan\alpha \in \mathbb{Q} , \ \gamma \in \mathbb{C}^1 \quad , \qquad (4.11)$$

where γ is a non vanishing solution of the corresponding supporting polynomial which will be analysed now (compare Appendix A.1).

Assume the straight line of the segment I considered meets exactly p points of the Exponent Diagram corresponding to the coefficients

$$a_{j_l}^{i_l} \quad (l = 1 , \ldots, p; \ \ j_1 < \ldots < j_p ; \ \ i_1 > \ldots > i_p) . \qquad (4.12)$$

This straight line is associated by a polynomial of degree $(j_p - j_1)$ of the form

$$0 = \sum_{l=1}^{p} a_{j_l}^{i_l} . k^{i_l} . s^{j_l} \quad , \qquad (4.13)$$

where $(j_p - j_1)$ is equal to the length of the projection of this straight line onto the $e(s)$ - axis.
Using the relations

$$\nu_l := j_l - j_1, \ \nu := j_p - j_1 \quad ,$$

$$\mu_l := i_l - i_p, \ \mu := i_1 - i_p \quad , \quad (\mu_p = 0, \ \nu_1 = 0)$$

and (4.14)

$$b_l := a_{j_l}^{i_l} , \ \beta = \frac{\mu_1}{\nu_p} = \frac{\mu}{\nu} , \ l \in \{1, \ldots , p\} \quad ,$$

the equation (4.13) takes the form

$$o = \sum_{l=1}^{p} b_l \cdot k^{\mu_l} \cdot s^{\nu_l} \quad \text{for} \quad s \neq o \, , \, k \neq o. \tag{4.15}$$

Inserting (4.11) into (4.15) provides the expression

$$o = \sum_{l=1}^{p} b_l \cdot \gamma^{\nu_l} \cdot k^{\mu_l + ß.\nu_l} \quad . \tag{4.16}$$

Taking into account the relation

$$\mu_1 + ß.\nu_1 = \ldots = \mu_p + ß.\, \nu_p \quad , \tag{4.17}$$

equation (4.16) is for $k \neq o$ equivalent to the corresponding <u>supporting polynomial</u>

$$o = \sum_{l=1}^{p} b_l \cdot \gamma^{\nu_l} \tag{4.18}$$

as a conditional equation for the $(j_p - j_1) = \nu_p$ non vanishing constants γ .

This procedure has to be performed for each straight line of the segment I and -as will be seen later on- for each of the segments II to VIII of the Exponent Diagram.
In case of a concrete computation of the small solutions of (4.3) it is useful to consider the following two situations :

<u>Case 1 :</u> The straight line of the segment I meets exactly two points of the Exponent Diagram, corresponding to the coefficients a_j^i and a_m^l , where $j < m$ and $i > l$.
Then the equation (4.13) takes the simplified form

$$o = a_j^i . k^i . s^j + a_m^l \cdot k^l \cdot s^m \quad , \tag{4.19}$$

which for

$$k \neq o, \; s \neq o, \; b := a_j^i \neq o \; , \; g := a_m^l \neq o \; , \; \mu := i - l \; \text{ and } \; \nu := m - j \tag{4.20}$$

takes the form

$$o = b.k^{\mu} + g.\, s^{\nu} \quad . \tag{4.21}$$

Inserting (4.11) into (4.21) delivers the relation

$$o = b.k^{\mu} + g.\, \gamma^{\nu} . k^{ß.\nu} \quad ,$$

or using

$$ß := \frac{\mu}{\nu} \quad , \quad \mu = ß.\nu \quad \text{and} \quad k \neq 0 \tag{4.22}$$

implies the equations

$$o = b + g.\gamma^{\nu} \quad \text{and} \quad o = 1 + \frac{g}{b}.\gamma^{\nu} \quad . \tag{4.23}$$

Then the solutions of (4.19) and of (4.23) take the form

$$s = (- \frac{b}{g}. k^{\mu})^{1/\nu} \ , \text{ where } \quad \gamma := (- \frac{b}{g})^{1/\nu} \quad . \tag{4.24}$$

Because of its importance in electrical network theory, the equation (4.23) is called "Butterworth Polynomial" [4.17] . The set of the zeros of (4.23) are said to create a "Butterworth Configuration". Transfer functions of linear electrical filters and control systems with no zeros and with poles which constitute a Butterworth Configuration ensure a distinguished transient behaviour of these systems.

Case 2 : If a straight line of the segment I of the Exponent Diagram meets more than two points, then the constant γ has to be computed numerically from the supporting polynomial (4.18). The corresponding roots will no longer build a Butterworth Polynomial.

Note 9 : (The supporting polynomial).

From now on, the supporting polynomial corresponding to a straight line of any one of the segments I, IV, V and VIII of the Exponent Polygon will be constructed as follows :

(i) All terms of the equation (4.2) which don't correspond to a point on this straight line are omitted.

(ii) In the residual of the terms, the factors s^j are replaced by the factors γ^j, and the factors k^i are omitted.

(iii) The non vanishing roots of the resultant polynomial are the roots of the supporting polynomial of the straight line. They constitute the factors γ_j of the different small, mediumsized and large solutions of (4.2), the structure of which will be investigated now.

The statement of this note is easily proved by using the transformation relations (4.7) and (4.8) in explicit form.

STEP 1o : Angles of departure at the point $s = o$ for $k = o$ (segment I).

In analogy to step 9, two cases are treated.

Case 1 : The straight line of the segment I meets exactly two points of the Exponent Diagram corresponding to the coefficients $a_j^i =: b_1$ and $a_m^1 =: b_2$, where $j < m$, $i > 1$, $\mu := i-1$ and $\nu := m-j$. Then in agreement with step 9, case 1, (compare equation (4.21)), exactly ν root locus branches leave the point $s = o$ for $k > o$. The corresponding angles of departure are determined by the relation

$$\varphi_\lambda(o,o) := \varphi_\lambda(s,k)\Big|_{s=o,\,k=o} = \frac{1}{\nu}\cdot\{\arg(b) - \arg(g) + \mu\cdot\arg(k) + (2\cdot\lambda - 1)\cdot\pi\}$$

for $\lambda = 1, \ldots, \nu$ or (4.25a)

$$\varphi(o,o) = \begin{cases} \frac{1}{\nu}\arg\left(\frac{b}{g}\right) + (2\cdot\lambda - 1)\cdot\pi & \text{for } \mu = 2\cdot m \text{ resp. for } k > o;\ m \in \mathbb{N}, \\ \frac{1}{\nu}\arg\left(\frac{b}{g}\right) + 2\cdot\lambda\cdot\pi & \text{for } \mu = 2m+1 \text{ and for } k < o\,. \end{cases}$$

Case 2 : The straight line of the segment I of the Exponent Polygon considered, meets at least two points of the Exponent Diagram. Then the corresponding angles of departure of root locus branches from the point $s = o$ are determined by the relation.

$$\varphi_\lambda(o,o) = \arg(\gamma_\lambda) + \beta\cdot\arg(k) = \arg(\gamma_\lambda) + \frac{\mu}{\nu}\cdot\arg(k) \quad (4.25b)$$

for $\lambda = 1, 2, \ldots, \nu$; $\beta = \mu/\nu$,

where the constants γ_λ are the non vanishing roots of one of the supporting polynomials of segment I of (4.2), and the constant β is defined by a straight line of this segment. This procedure has to be repeated for each straight line of the segment I.

Proof of step 1o :

According to step 2, the behaviour of the root locus branches near the point (s,k) = (o,o) is completely determined by segment I of the Exponent Polygon. Therefore, the discussion in connection with step 9 and with note 9 proves the statements of step 1o. In agreement with step 4, there are e_I branches of this type.

Simple branching diagrams in connection with the corresponding angles of departure of case 1 of step 1o are sketched in Figures 4.2a and 4.2b.

Note 1o :

The statements of the following steps 11 to 19 are direct consequences of step 1o, taking into account the transformation relations (4.7) and (4.8).

STEP 11 : Angles of arrival at a sink s = o for |k| = ∞ (segment V of (4.2)).

In analogy to step 1o, it is useful to treat two cases.

Case 1 : The straight line of the segment V considered meets exactly two points of the Exponent Diagram. Then the angles of arrival at s = o for k = ∞ are determined by the relation

$$\varphi_\lambda(o,\infty) := \varphi_\lambda(s,k)\Big|_{s=o,k=\infty} = \frac{1}{\nu}\cdot\left\{\arg\left(\frac{b}{g}\right) - \mu\cdot\arg(k) + (2\cdot\lambda - 1)\cdot\pi\right\} + \pi , \tag{4.26a}$$

where in agreement with the relations (3.12b) μ is defined as

$$\mu := i' - l' := l - i \quad , \quad i' := q - l \quad ,$$

and the constants ν , ß, b and g are defined by the relations (4.11), (4.19) and (4.2o), corresponding to the segment V. The case 1 always provides a Butterworth Configuration.

Case 2 : The straight line of the segment V considered meets

$b.g < o \quad , \quad b.k^{\mu} + g.s^{\nu} = o$

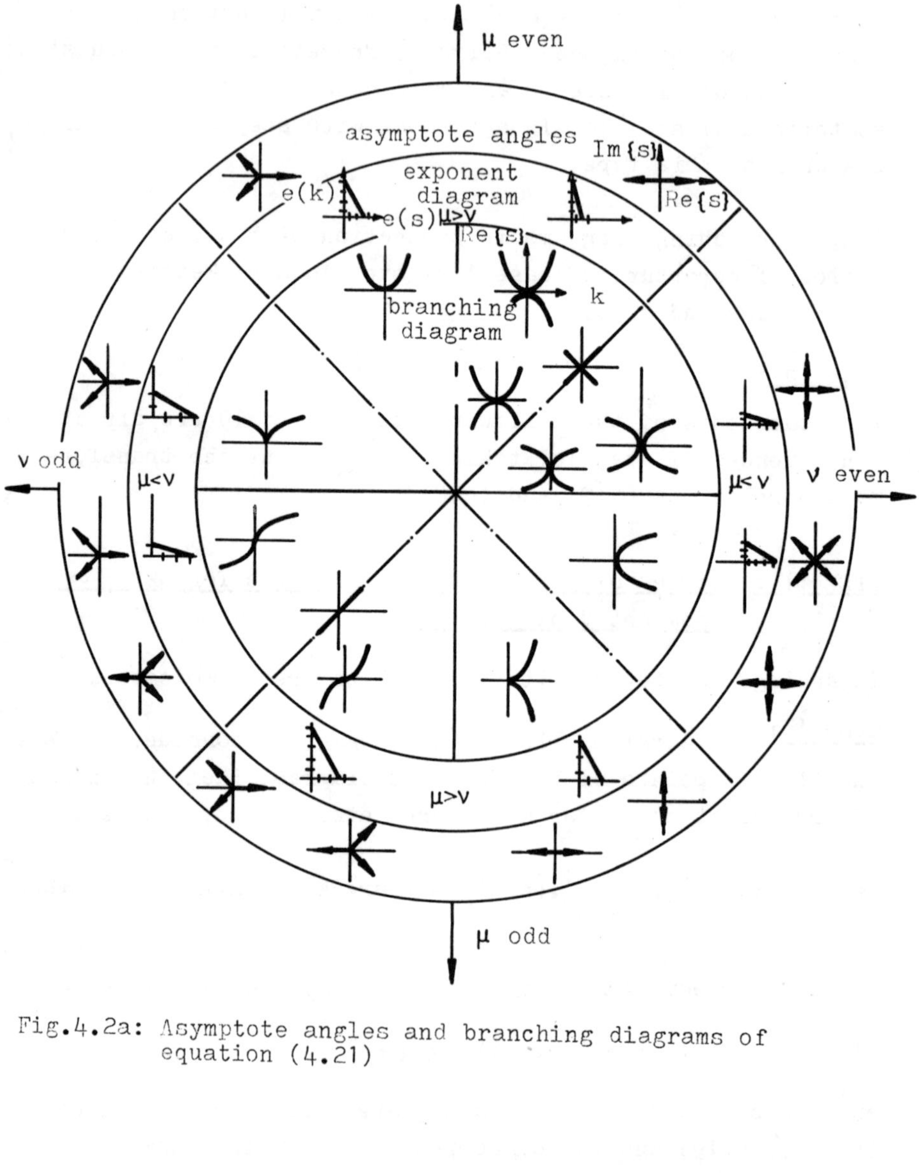

Fig.4.2a: Asymptote angles and branching diagrams of equation (4.21)

$$b.g > o \quad , \quad b.k^{\mu} + g.s^{\nu} = o$$

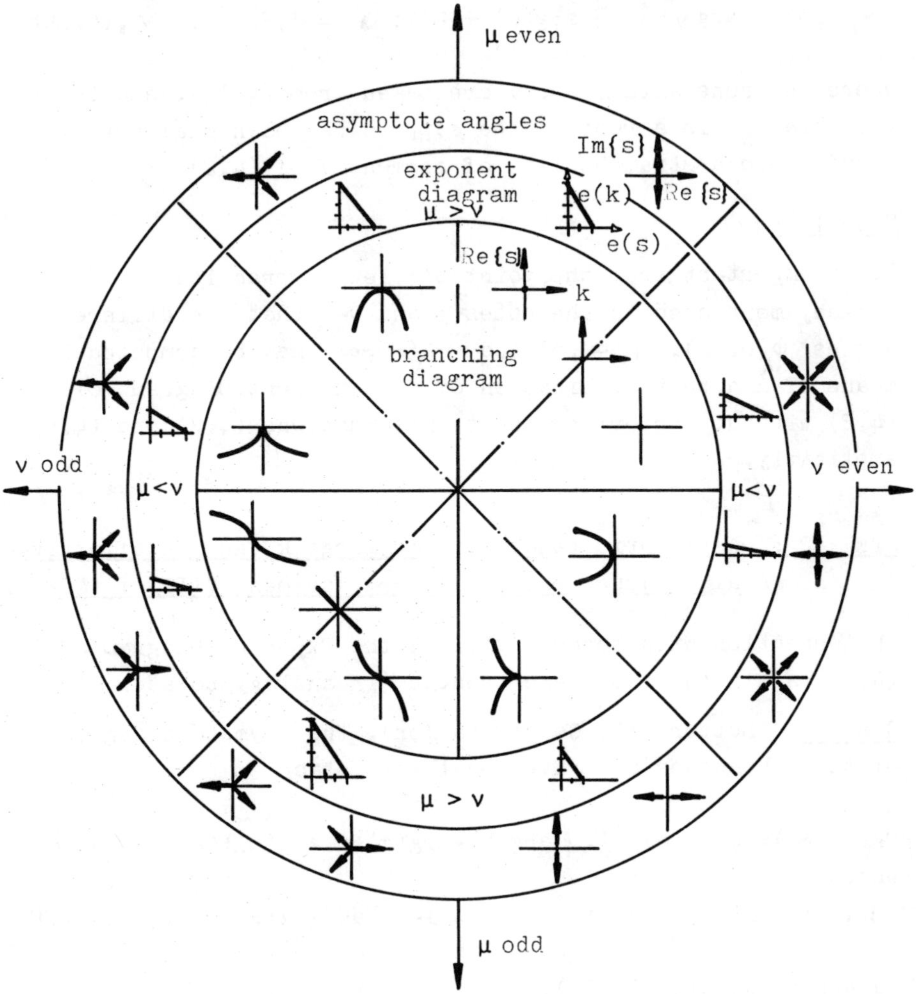

Fig.4.2b: Asymptote angles and branching diagrams of equation (4.21)

two or more than two points of the Exponent Diagram. Then the corresponding angles of arrival at $s = o$ for $k = \infty$ are determined by the relation

$$\varphi_\lambda(o,\infty) = \arg\gamma_\lambda - \frac{\mu}{\nu}.\arg(k) + \pi \quad ; \quad \lambda = 1,2, \dots ,\nu \ , \quad (4.26b)$$

where the constants μ and ν are taken from (4.14), and the variable γ_λ is a root of the supporting polynomial corresponding to a straight line of segment V of (4.2).

Note 11 :

It is important from the point of view of practice -as already mentioned in the notes 3 and 9- that the different angles $\varphi_\lambda(o,\infty)$, $\varphi_\lambda(\infty,o)$ and $\varphi_\lambda(\infty,\infty)$ may be computed by means of a direct application of the Exponent Diagram of (4.2) without performing the transformations (4.7) and (4.8) explicitly.

STEP 12 : Asymptotic behaviour for large k (angles of arrival at $|s| = \infty$ for $|k| = \infty$; segment VIII of (4.2)).

In connection with segment VIII of the Exponent Diagram of (4.2), again two cases are treated (in analogy to step 9):

Case 1: (Butterworth Configuration). The asymptote angles of the ν branches are given by the relation

$$\varphi_\lambda(\infty,\infty) = \varphi_\lambda(o,o) = \frac{1}{\nu}.\{\arg(b) - \arg(g) + \mu.\arg(k) + (2.\lambda - 1).\pi\} ,$$

where (4.27a)

$$\mu := i' - l' \ , \quad i' > l' \ , \quad i' := i-1 \quad \text{and} \quad \nu := |m-j| = |j'-m'|$$

in agreement with (4.20).

Case 2 : (No Butterworth Configuration). The asymptote angles are determined by the relation

$$\varphi_\lambda(\infty,\infty) = \varphi_\lambda(o,o) = \arg(\gamma_\lambda) + \beta.\arg(k)$$

$$= \arg(\gamma_\lambda) + \frac{\mu}{\nu}.\arg(k) \quad , \quad (4.27b)$$

where $\beta = \tan\alpha_{VIII.j} = \frac{\mu}{\nu}$ is taken from the segment VIII of (4.2) and γ is one of the ν roots of the supporting polynomial of this segment.

STEP 13 : Asymptotic behaviour for k=o (angles of arrival at $|s| = \infty$ for $|k| = o$; segment IV of (4.2)).

Case 1 : (Butterworth Configuration) The asymptote angles of the ν branches for k=o are determined by the relation

$$\varphi_\lambda(\infty, o)=\frac{1}{\nu}.(\arg b-\arg g-\mu.\arg k-(2.\lambda-1).\pi)+\pi, \lambda = 1,2,...,\nu, \tag{4.28a}$$

where the constants ν,μ, b and g are defined according to (4.21). They are to be taken from the segment IV of (4.2).

Case 2 : (No Butterworth Configuration) The ν asymptote angles are given by the relation

$$\varphi_\lambda(\infty, o) = \arg\gamma_\lambda - \frac{\mu}{\nu}.\arg k + \pi, \tag{4.28b}$$

where the constants $\beta = \tan\alpha_{IV} = \frac{\mu}{\nu}$ are determined by the segment IV of (4.2) and γ_λ is a non vanishing root of the supporting polynomial, corresponding to this segment.

STEP 14 : Angles of departure at the point $s_j^o \neq o$ for $k = o, j = 1,..., e_{II}$ (segment II).

The angles of departure $\varphi_\lambda(s_j^o, o)$ of root locus branches from a root $s_j^o \neq o$ of the polynomial $Q_o(s)$ are given by the relations (4.25a) and (4.25b) after having transformed the polynomial P(s,k) by means of the relation (4.8a) to the polynomial $P(s-s_j^o,k)$. The various constants of (4.25) are related to the segment I of $P(s-s_j^o,k)$.

STEP 15 : Angles of arrival at a finite sink $s_j^q \neq o$ for $|k| = \infty, j = 1,..., e_{VII}$ (segment VII).

The angles of arrival $\varphi_\lambda(s_j^q, \infty)$ of root locus branches at a finite non vanishing sink s_j^q of the polynomial $Q_q(s)$ are given by the relations (4.26) by transforming the polynomial $P(s,k)$ by means of the relation (4.8f) to the polynomial $P(s-s_j^q,k)$. The different constants of (4.26) are related to the segment V of $P(s-s_j^q,k)$.

Note 12 :

In analogy to the notes 3 and 9, the transformations (4.8) have not to be applied explicitely to the polynomial $P(s,k)$ to get the interesting terms of the polynomials $P_{II}(s-s_j^o,k)$ and $P_{VII}(s-s_j^q,k')$ (compare the examples of the following chapters).

Note 13 :

The case 1 of each of the steps 1o to 15 may be reformulated as follows :

Let $s_k^o \neq o$ be a l_i- fold root of the polynomial $Q_o(s)$. The angles of departure of root locus branches from the point s_k^o for $k = o$ may be computed in agreement with step 14, expression (4.25a), using the transformation (4.8a). Then we have

$$P_{II}(s',k) := P(s-s_k^o,k) = \sum_{i=1}^{q} k^i \cdot \sum_{j=1}^{n} {}_{II}Q_j(s') \quad , \qquad (4.29)$$

where

$$s' := s-s_k^o \quad , \quad {}_{II}Q_i\,(s') := s'^{\,l_i} \cdot {}_{II}\tilde{Q}_i(s')$$

or

$$(4.30)$$

$$ {}_{II}Q_i\,(s') = s'^{\,l_i} \cdot \sum_{j=o}^{n_i-l_i} {}_{II}a_j^i \cdot (s-s_k^o)^j$$

and

$$ {}_{II}Q_i(s') = s'^{\,l_i} \cdot {}_{II}\, a^i_{n_i-l_i} \cdot \prod_{\substack{j=1 \\ s_j^i \neq s_k^o}}^{n_i} (s_k^o - s_j^i) = s'^{\,l_i} \cdot a^i_{n_i-l_i} \quad , \qquad (4.31)$$

where

$$\text{degree } {}_{II}Q_i = \text{degree}\left({}_{II}\tilde{Q}_i\right) - l_i = n_i - l_i \quad ,$$

and (4.32)

$$a^{i}_{n_i-l_i} := \text{II}\, a^{i}_{n_i-l_i} \cdot \prod_{\substack{\varkappa_i=1 \\ s^i_\varkappa = s^o_k}}^{n_i} (s^o_k - s^i_\varkappa) \; .$$

After renaming the index $(n_i - l_i)$ into j , the coefficient a^i_j of (4.32) may be identified with one of the coefficients a^i_j of the equations (4.13) and (4.19). In case 1 (exactly two points of the Exponent Diagram are met by the Segment I of the corresponding Exponent Polygon of $P_{II}(s',k)$) we have for $i<j$ according to (4.25a) :

$$\varphi_\lambda(s^o_k, o) = \frac{1}{l_i - l_j} \cdot ((j-i).\arg(k) + \arg(a^j_{n_j-l_j}) - \arg(a^i_{n_i-l_i})$$

$$+ (2.\lambda-1).\pi + \sum_{\varkappa=1,\; s^i_\varkappa \neq s^o_k}^{n_j} \arg\;(s^o_k - s^i_\varkappa) - \sum_{\varkappa=1,\; s^j \neq s^o_k}^{n_i} \arg(s^o_k - s^j_\varkappa)) \tag{4.33}$$

Inserting the relations

$$\begin{aligned} & q = 1 \;, \; r_1 = 1 \;, \\ & n_o = n \;, \; n_1 = m \leq n \;, \\ & a^o_j = a_j \quad \text{for} \quad j = 0,1,\dots,n \;, \quad s^o_j =: s_{p_j} \quad \text{for} \quad j=1,\dots,n, \\ & a^1_i = b_i \quad \text{for} \quad i = 0,1,\dots, m \quad \text{and} \quad s^1_i = s_{n_j} \quad \text{for} \quad i=1,\dots,m \end{aligned} \tag{4.34}$$

into the equations (4.3) to (4.6) results in the characteristic equation of a single loop linear control system

$$o = \sum_{j=0}^{n} a_j . s + k \cdot \sum_{i=0}^{m} b_i . s^i = a_n . \prod_{j=1}^{n} (s + s_{p_j}) + k.b_m . \prod_{i=m}^{m} (s + s_{n_i}) \;, \tag{4.35}$$

which is identical to equation (4.1). The well known corresponding angles of departure of root locus branches from the pole s_{p_k} take the form

$$\varphi_\lambda(s_{p_k},o)=\frac{1}{\varrho-\varkappa}\cdot\left\{\arg(k.\frac{a_n}{b_m})+\sum_{\substack{j=1\\j\neq k}}^{n}\arg(s_{p_k}+s_{p_j})-\sum_{i=1}^{m'}\arg(s_{p_k}+s_{n_i})+(2.\lambda-1).\pi\right\},$$ (4.36)

where $m'=m-\varkappa$, $\varkappa$ is the multiplicity of the pole s_{p_k} and $\lambda\in\{1,\ldots,\varrho-\varkappa\}$.

A comparison of the relations (4.36) and (4.33) shows that the angles of departure of root locus branches from a pole of the open loop transfer function of a single loop control system and of a multiloop control system may be computed from the same formula in case 1. The same result holds for the computation of the angles of arrival at a sink s_j^q (or of a zero) as well as for the computation of the asymptote angles of root locus branches for large real values of k ($k>0$).

STEP 16 : Angles of departure from s = o for non vanishing finite values of k (segment III).

The angles of departure $\varphi_\lambda(o,k_{1_o}^j)$ of root locus branches from the point s = o for non vanishing finite real values $k_{1_o}^j$ are computed from the relations (4.25) on the basis of the segment I of the transformed polynomial $P_{III}(s,k')$, which follows from $P(s,k)$ by means of the transformation (4.8b) where $k_{1_o}^j$ is a real non vanishing root of the polynomial $R_{1_o}(k)$.

STEP 17 : Asymptote angles for non zero finite real values of k (segment VI).

The asymptote angles $\varphi_\lambda(\infty,k_n^j)$ of root locus branches for $k_n^j\in\mathbb{R}^1$, $0<|k_n^j|<\infty$, where k_n^j is a root of $R_n(k)$, are determined by the relations (4.27a) and (4.27b) on the basis of the polynomial $P_{VI}(s',k')$ which follows from $P(s;k)$ by means of the transformation (4.8e).

Note 14 :

Again (compare note 12), only the coefficients of $P_{III}(s,k')$ and of $P_{VI}(s',k')$ defining a straight line of the segment I have to be computed explicitly by means of the relations (4.8b) and (4.8e).

STEP 18 : Asymptote points (segment VIII).

Asymptote points are points of the complex plane which are met by at least one of the asymptotes of the root locus branches. In connection with the asymptote angles they uniquely determine the asymptotes of the root locus branches. Because of step 6 (symmetry of locus), asymptote points either are real or they appear as complex conjugate pairs.
The asymptote points corresponding to a straight line of the segment VIII of P(s,k) may be computed according to the following process (compare Fig. 4.3).

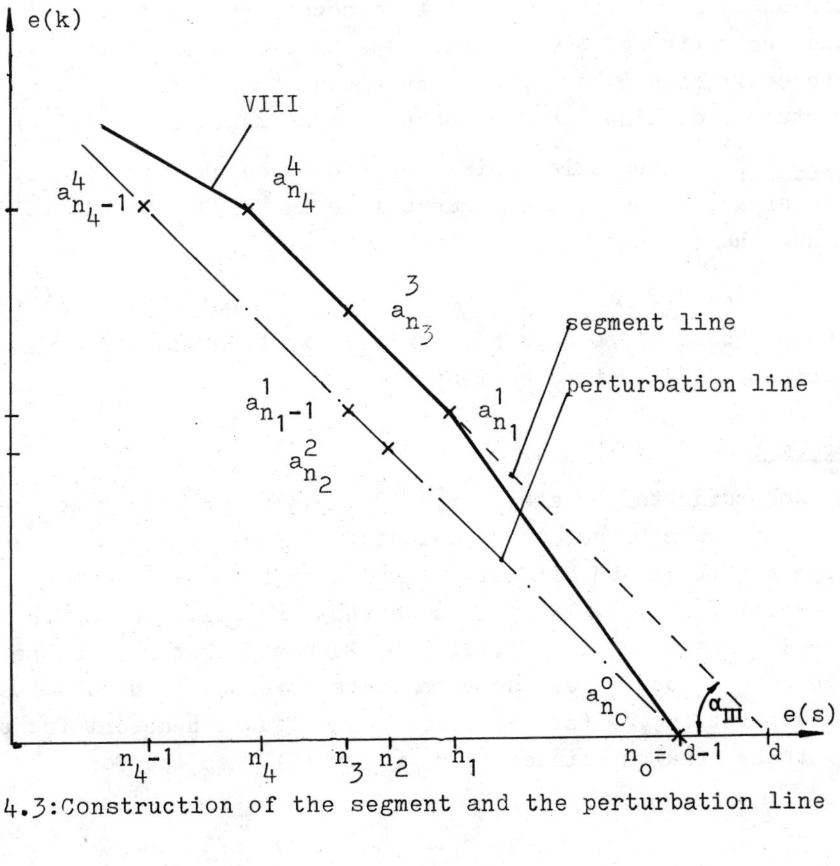

Fig.4.3:Construction of the segment and the perturbation line

Extending the straight line of segment VIII considered to its intersection point (d,o) with the abscissa e(s), and drawing a straight line through the point (d-1,o) in parallel to the first one, the latter straight line will be called "perturbation line" whereas the first straight line will be called "segment line" (cf. Fig. 4.3).
Then two cases have to be considered :

Case a : There are no points inside the area limited by the segment line and by the perturbation line. Then the asymptote points s_{o1} are determined by the relation

$$s_{o1} = \frac{\sum\limits_{j \in J} a^j_{n_j} \cdot \gamma_1^{n_j}}{\psi(\gamma_1)} , \qquad (4.37a)$$

where γ_1 is a root of the supporting polynomial $\phi(\gamma)$ corresponding to the segment line considered, and $\psi(\gamma_1)$ is the derivative of $\phi(\gamma)$ with respect to γ at the point γ_1. The coefficients $a^j_{n_j}$, $j \in J$ correspond to points (n_j, j) on the perturbation line defined by the segment line considered.

Case b : There exist points of the Exponent Diagram inside the area limited by the segment line and by the perturbation line. Then we have

$$|s_o| = \infty \ . \qquad (4.37b)$$

In this case there does not exist a finite asymptote point corresponding to this segment line.

Proof :

a) According to the steps 5,8,12 and 13, the asymptote angles of root locus branches are computed on the basis of the segments IV,VI and VIII of the Exponent Diagram of the polynomial P(s,k). The points on these segments belong to the coefficients $a^i_{n_i}$, i=1, ... ,p. Each straight line corresponding to one of these segments intersects the abscissa at a point $(d_\varkappa, o)$ (cf. Fig. 4.4). The equation describing each of these straight lines (for $\alpha_\varkappa \neq \pi/2$) has the form

$$e(s) = d_\varkappa (\mp) \frac{1}{\beta_\varkappa} \cdot e(k) \quad , \text{ where } \quad \beta_\varkappa = \tan \alpha_\varkappa \quad . \tag{4.38}$$

The computation of an asymptote point s_0 corresponding to a straight line of the segment VIII is based on the points upon the <u>segment line (4.38)</u> as well as on points upon the <u>perturbation line</u> which is described by the equation

$$e(s) = (d_\varkappa - 1) - \frac{1}{\beta_\varkappa} \cdot e(k) \quad . \tag{4.39}$$

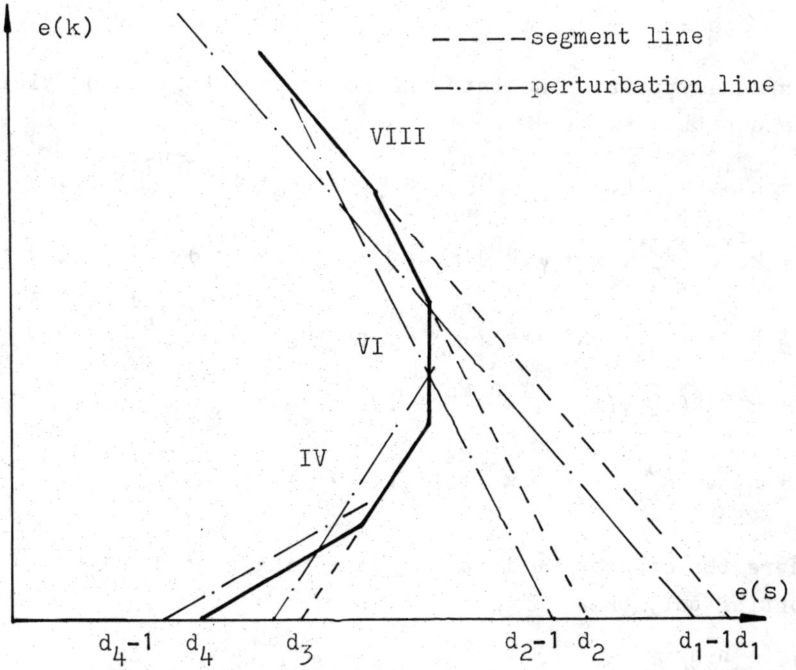

Fig.4.4: Segment line and perturbation line corresponding to segments IV and VIII .

The coefficients of the points on the perturbation line are $a^j_{n_j}$. For the sake of simplicity, the coefficients $a^i_{n_i}$ corresponding to points on the segment line are chosen as

$$a^1_{n_1} \quad , \quad a^3_{n_3} \quad \text{and} \quad a^4_{n_4} \quad ,$$

where

$$n_0 > n_1 > n_2 > n_3 > n_4 \quad ,$$

whereas the coefficients corresponding to points upon the perturbation line are chosen as

$$a^o_{n_o} \quad , \quad a^1_{n_1-1} \quad , \quad a^2_{n_2} \quad \text{and} \quad a^4_{n_4-1} \quad , \quad \text{i.e.} J = \{ 0,1,2,4 \}$$

(cf. Fig. 4.3).
Then inserting the ansatz

$$s = s_0 + \gamma . k^{ß}$$

for an asymptote line into the relation $P(s,k) = 0$ yields the equation

$$\begin{aligned} o = k^{r_1} . (a^1_{n_1} . (s_0 + \gamma . k^{ß})^{n_1} + a^1_{n_1-1} . (s_0 + \gamma . k^{ß})^{n_1-1} + \ldots) \\ + k^{r_3} . (a^3_{n_3} . (s_0 + \gamma . k^{ß})^{n_3} + a^3_{n_3-1} . (s_0 + \gamma . k^{ß})^{n_3-1} + \ldots) \\ + k^{r_4} . (a^4_{n_4} . (s_0 + \gamma . k^{ß})^{n_4} + a^4_{n_4-1} . (s_0 + \gamma . k^{ß})^{n_4-1} + \ldots) \\ + k^{r_2} . (a^2_{n_2} . (s_0 + \gamma . k^{ß})^{n_4} + \ldots) \\ + k^{0} . (a^o_{n_o} . (s_0 + \gamma . k^{ß})^{n_o} + \ldots) \quad , \end{aligned}$$

where the constant γ is a non vanishing root of the supporting polynomial

$$a^1_{n_1} . \gamma^{n_1-n_4} + a^3_{n_3} . \gamma^{n_3-n_4} + a^4_{n_4} \quad \text{and} \quad ß = \tan \alpha_{VIII} .$$

Comparing coefficients of terms with equal exponents, and taking into account the equations

$$n_1 . ß + r_1 = n_3 . ß + r_3 = n_4 . ß + r_4$$

and

$$n_o . ß = (n_1 - 1) . ß + r_1 = n_2 . ß + r_2 = (n_4 - 1) . ß + r_4$$

provides the relation

$$0 = (a^1_{n_1} \cdot \gamma^{n_1} + a^3_{n_3} \cdot \gamma^{n_3} + a^4_{n_4} \cdot \gamma^{n_4}) . k^{n_1 \cdot \beta + r_1}$$

$$+ \{ s_0 \cdot (n_1 \cdot a^1_{n_1} \cdot {}^{n_1-1} + n_3 \cdot a^3_{n_3} \cdot {}^{n_3-1} + n_4 \cdot a^4_{n_4} \cdot {}^{n_4-1})$$

$$+ (a^o_{n_o} \cdot \gamma^{n_0} + a^1_{n_1-1} \gamma^{n_1-1} + a^2_{n_2} \cdot \gamma^{n_2} + a^4_{n_4-1} \cdot \gamma^{n_4-1}) \} . k^{(n_1-1) \cdot \beta + r_1} + \dots .$$

In general we have for $\gamma \neq 0$ and for great values of k :

$$- s_0 = \frac{\sum_{j \in J} a^j_{n_j} \cdot \gamma_1^{n_j}}{\sum_{i \in I} n_i \cdot a^i_{n_i} \cdot \gamma^{n_i - 1}} ,$$

where the index pairs (n_i, i) correspond to points on the segment line (in the above case $(n_1, 1)$, $(n_3, 3)$, $(n_4, 4)$) and the pairs (n_j, j) correspond to points on the perturbation line (in the above case $(n_1-1, 1)$, $(n_4-1, 4)$, $(n_0, 0)$ and $(n_2, 2)$) .

Note 15 : (Special cases of step 18)

a. Assume that in case a of step 18 no points of the Exponent Diagram are met by the perturbation line. Then we have $s_{0j} = 0$ $(J = \phi)$.

b. Assume that in case a of step 18 only two points of the Exponent Diagram corresponding to the coefficients $a^i_{n_i}$ and $a^j_{n_j}$ $(n_i < n_j)$ are met by the segment line and only two points corresponding to the coefficients $a^i_{n_i-1}$ and $a^j_{n_j-1}$ are met by the perturbation line.

Then we have

$$s_0 = - \frac{a^j_{n_j-1} \cdot \gamma^{n_j - n_i} + a^i_{n_i-1}}{n_j \cdot a^j_{n_j} \cdot \gamma^{n_j - n_i} + n_i \cdot a^i_{n_i}} .$$

The relation

$$\gamma^{n_j - n_i} \cdot a^1_{n_j} + a^i_{n_i} = o$$

provides the equation

$$s_o = - \frac{a^j_{n_j-1} \cdot \left(- \frac{a^i_{n_i}}{a^j_{n_j}}\right) + a^i_{n_i-1}}{n_j \cdot a^j_{n_j} \cdot \left(- \frac{a^i_{n_i}}{a^j_{n_j}}\right) + n_i \cdot a^i_{n_i}}$$

and

$$s_o = - \frac{\frac{a^j_{n_j-1}}{a^j_{n_j}} - \frac{a^i_{n_i-1}}{a^i_{n_i}}}{n_j - n_i} ,$$

respectively. Using the abbreviations

$$\text{trace}\{Q_i\} := - \frac{a^i_{n_i-1}}{a^i_{n_i}} \quad \text{and} \quad \text{trace}\{Q_j\} := - \frac{a^j_{n_j-1}}{a^j_{n_j}}$$

yields the final relation

$$s_o = \frac{\text{trace}\{Q_j\} - \text{trace}\{Q_i\}}{n_j - n_i} , \qquad (4.40)$$

which is the well known formula of <u>classical root locus</u> techniques for asymptote points.

<u>Note 16 :</u>

a. The asymptote points corresponding to a straight line of segment IV of P(s,k) may be computed by replacing the polynomial P(s,k) by means of the transformed polynomial $P_{IV}(s,k')$, where $k' := 1/k$, and by applying the formula (4.37) to the corresponding segment VIII of $P_{IV}(s,k')$.

b. The asymptote points corresponding to a straight line of the segment VI of P(s,k) may be computed by replacing the polynomial P(s,k) by means of the transformed polynomial $P_{VI}(s,k')$, where $k' := 1/(k-k_n^j)$, where k_n^j is one of the real non vanishing finite roots of the polynomial $R_n(k)$, and by applying the formula (4.37) to the corresponding segment VIII.

The following two steps are trivial consequences of the division theory of classical polynomial algebra. They are presented here without proof.

STEP 19 : Breakaway or reentry points.

The breakaway or reentry points (s,k) of root locus branches from or towards the real axis are simultaneous roots of the polynomials

$$P(s,k) \quad \text{and} \quad \partial P(s,k)/\partial s \ . \tag{4.41a}$$

According to well known results from classical algebra, these points may be computed as roots of the resultant

$$R\ (P(s,k),\ \partial P(s,k)/\partial s)\ . \tag{4.41b}$$

STEP 2o : Return points.

The return points of root locus branches on the real axis are among the common real roots of the polynomials

$$P(s,k) \quad \text{and} \quad \partial P/\partial k\ . \tag{4.42a}$$

These points may be computed as real roots of the resultant

$$R\ (P(s,k),\ \partial P(s,k)\ /\partial k)\ . \tag{4.42b}$$

In the simple cases for q=2 and q=3 we have

$$R(P,P_k') = Q_2 \cdot (Q_1^2 - 4 \cdot Q_2 \cdot Q_0) \quad \text{for} \quad (q=2)\ , \text{and} \tag{4.43}$$

$$R(P,P_k') = Q_3 \cdot (Q_0 \cdot (18 \cdot Q_1 \cdot Q_2 \cdot Q_3 - 27 \cdot Q_0 \cdot Q_3^2 - 4 \cdot Q_2^3) + Q_1^2 \cdot (Q_2^2 - 4 \cdot Q_1 \cdot Q_3))$$

for q=3, respectively.

Note 17 :

Step 2o implies that first order root locus plots corresponding to equation (4.1) can not produce return points on the real axis of the complex plane. The corresponding resultant (4.42b) is not defined. As is easily seen, the existence of return points is closely related to the concept of the complete or partial pole sink compensation as described in note 7. Partial pole sink compensation occurs if some of the polynomials $Q_j(s)$ corresponding to distinguished j have a common root s_i^j, whereas s_i^j is not a common root of all of the polynomials $Q_j(s)$, $j = 1, \dots ,q$.
As is easily seen, the resultant (4.42b) vanishes at the sinks s_j^q of the system. The sinks are trivial return points. Moreover, for q=2 the resultant (4.42b) vanishes for $Q_2(s) = 0$ and for common roots of the polynomials $Q_0(s)$ and $Q_1(s)$. For q=3, the resultant vanishes for $Q_3(s)=0$ and for the common roots of $Q_0(s)$ and of $Q_1(s)$.

Note 18 : Behaviour along the real axis.

The existence of return points of root locus branches along the real axis indicates that there exist parts of the real axis which are covered by more than one root locus branch and there exist segments of the real axis which are not covered at all by root locus branches for $-\infty \leq k \leq +\infty$, both contrary to the classical root locus plots. Therefore, rules which are as simple as the related rules of classical root locus technique can not be expected. There exist a lot of special cases in connection with the higher order root locus technique, which may be characterized by a corresponding rule. Two of these rules are formulated by the following two steps (compare [4.3]).

STEP 21 : Sum of the closed-loop poles.

Let $n=n_o \geq (n_i+2)$ for $i>0$. Then we have :

a) The ratio $\frac{R_{n_o-1}}{R_{n_o}}$ is independent of k.

b) The sum of the real parts of the roots of the closed loop $(s_j(k))$ is independent of k, i.e.

$$\sum_{j=1}^{n=n_o} \mathrm{Re}\{s_j(k)\} = \sum_{j=1}^{n_o} \mathrm{Re}\{s_j^o\} \tag{4.44}$$

Proof :

a) The relation $n_o=n \geq n_i+2$ for $i>0$ implies that the expression $R_{n_o}(k)$ is a non vanishing constant, and that the expression $R_{n_o-1}(k)$ vanishes identically in k. This proves part a) of the statement (compare Fig. 3.2b).

b) From equation (4.6) in connection with part a) we have (because of $R_{n_o} \neq 0$) :

$$\frac{P(s,k)}{R_{n_o}} = \frac{R_o(k)}{R_{n_o}} + \frac{R_1(k)}{R_{n_o}} \cdot s + \dots + \frac{R_{n_o-1}(k)}{R_{n_o}} \cdot s^{n_o-1} + s^{n_o} . \tag{4.45}$$

The statement b) in connection with statement a) provide the relation

$$\mathrm{trace}\left\{\frac{P(s,k)}{R_{n_o}}\right\} := -\frac{R_{n_o-1}(k)}{R_{n_o}} = \sum_{j=1}^{n_o=n} s_j(k) = \sum_{j=1}^{n} \mathrm{Re}\{s_j(k)\} . \tag{4.46}$$

STEP 22 : Product of the closed-loop poles.

Let the ratio $\frac{R_o}{R_{n_o}}$ be independent of k (which according to step 21 is ensured by the relation $n_o \geq n_i + 2$ for $i>0$), and let R_o= constant $\neq 0$. Then the product of the closed loop poles $s_j(k)$ is independent of k. This is equivalent to the relation

$$\prod_{j=1}^{n_o} s_j(k) = \prod_{j=1}^{n_o} s_j^o \tag{4.47}$$

Proof :

This statement follows directly from the obvious relation

$$\frac{R_o(k)}{R_{n_o}} = \prod_{j=1}^{n_o = n} s_j(k)$$

in connection with the statement b) of step 21.

Rules of classical root locus technique similar to those of steps 21 and 22 are well known in literature.

3. Summary of the higher order root locus construction rules.

In what follows, a summary of root locus construction rules is collected. In concrete situations, in general, only a subset of these rules is needed.

1. Synopsis of the various polynomials $Q_o(s)$ to $Q_q(s)$ and $R_o(k)$ to $R_n(k)$.

2. Computation of the roots of some or all of the polynomials $Q_o(s)$ to $Q_q(s)$ and $R_o(k)$ to $R_n(k)$.

3. Sketch of the Exponent Diagram and of the Newton Diagram (comp. Fig. 3.1 and 3.2).

4. Marking of the various segments I to VIII and drawing of the angles $\alpha_{I.1}$ to $\alpha_{VIII.j}$ of the **Exponent Polygon** (Fig. 3.1).

5. Collection of the constants

   l_o , q , n , n_o , n_q , e_I , e_{II} , e_{IV} , e_V , e_{VII} , e_{VIII}, $\bar{e}_{III}$ and $\bar{e}_{VI}$

   of the **Exponent Diagram** (Fig. 3.1 and 3.2), where

a. The constant l_o determines the number of complete pole sink compensations at the point s=o; l_o is equal to the number of root locus branches which are degenerated to the point s=o.

b. e_I is the number of non degenerated root locus branches starting at s=o for k=o.

c. e_{II} is the number of root locus branches starting at non vanishing finite poles of the open loop system.

d. $n_o = l_o + e_I + e_{II}$ is the total number of finite poles of the open-loop system, $(l_o + e_I)$ of which are equal to zero.

e. e_{IV} is the number of root locus branches starting at infinity for k=o.

f. $n = n_o + e_{IV}$ is the total number of root locus branches, l_o of which are degenerated to the point s=o.

g. $\bar{e}_{III}$ is the upper limit of the number of root locus branches which meet the point s=o for non vanishing, finite, real values k^j_{lo} of k (k^j_{lo} is a real root of $R_{lo}(k)$).

h. $\bar{e}_{VI}$ is the upper limit of the number of root locus branches which tend to infinity for non vanishing finite real values k^j_n of k (k^j_n is a real root of $R_n(k)$).

i. e_V is the number of root locus branches which end at the point s=o for infinite real values of k ($l_o + e_V$ is the number of vanishing sinks of the system, l_o of which correspond to root locus branches degenerated to the point s=o).

j. e_{VII} is the number of finite non vanishing end points s^q_j (sinks) of root locus branches for $k=\infty$ (s^q_j are the non vanishing roots of the polynomial $Q^q(s)$).

k. $n_q = l_o + e_V + e_{VII}$ is the total number of finite end points (sinks) of root locus branches (l_o of which are degenerated to the point s=o).

l. e_{VIII} is the number of root locus branches which tend to infinity for $k=\infty$, k real.

m. The number of root locus branches degenerated to a point $s_j^o = s_j^q$ is equal to the multiplicity of s_j^o as a common root of all polynomials $Q_j(s)$, $j=0, 1, \dots, q$.

6. <u>(Symmetry of locus)</u>
The root locus diagram is symmetrical with respect to the real axis.

7. Computation of the variables

$$\beta_{I.1}, \beta_{I.2} \text{ to } \beta_{VIII.j} \text{ and } \beta_{I.1} := \tan\alpha_{I.1} = \frac{\mu_{I.1}}{\nu_{I.1}},$$

where the values of $\mu_{I.1}$, $\nu_{I.1}$, ... , are taken from the **Exponent Polygon** (comp. Fig. 3.1b).

8. <u>Drawing of the n_o zeros s_j^o</u> of the polynomial $Q_o(s)$ as points "x" into the complex plane using $Re\{s\}$ as abscissa and $Im\{s\}$ as ordinate. These n_o points s_j^o are the finite starting points of n_o root locus branches (degenerated or not, simple or multiple) for k=o (segment I).

9. <u>Inserting of the n_q sinks</u> (sinks of branches) s_j^q of the polynomial $Q_q(s)$ as points "o" into the complex plane. These n_q points are the finite goals of n_q of the root locus branches for $k=\infty$ (degenerated or non degenerated, multiple or simple; segment VII).

1o. <u>Angles of departure from the point s=o for k=o (segment I)</u>
Two different cases must be considered :

<u>Case 1 :</u> Exactly two points of the Exponent Diagram corresponding to terms $a_j^i.k^i.s^j$ and $a_m^l.k^l.s^m$ $(i>l, j<m)$ of (4.2) are met by the straight line of the Exponent Polygon considered. Then the angles of departure from s=o for k=o take the form

$$\varphi(o,o) = \frac{1}{\nu}.\left\{\arg\left(\frac{b}{g}\right) + \mu.\arg(k) + (2.\lambda - 1).\pi\right\} \qquad (4.25a)$$

for $\lambda = 1, \ldots, v$, where $b := a_j^i$, $g := a_m^l$ and

$\beta = \frac{\mu}{v}$ ($\mu := i - l$, $v := m - j$).

Case 2 : The straight line of the segment I considered meets at least two points of the Exponent Diagram. Then we have

$$\varphi_\lambda(o,o) = \arg \gamma_\lambda + \beta . \arg(k) \quad \text{for} \quad \lambda = 1, \ldots, v \quad , \qquad (4.25b)$$

where γ_λ is one of the v non vanishing roots of the supporting polynomial corresponding to this line, and β is taken from the **Exponent Polygon.**

11. Angles of departure from any of the e_{II} non vanishing roots s_j^o of $Q_o(s)$.
Rule 1o is applied to segment I of the transformed equation $P(s-s_j^o;k)$, where only those coefficients of $P(s-s_j^o;k)$ must be computed which define the supporting polynomial of segment I related to $P(s-s_j^o;k)$.

12. Angles of arrival at a sink $s_j^q = o$ for $|k| = \infty$.
In line to Rule 1o we have :

Case 1 :

$$\varphi_\lambda(o,\infty) = \frac{1}{v} . \left\{ \arg(\tfrac{b}{g}) - \mu . \arg(k) + (2.\lambda - 1).\pi \right\} + \pi \qquad (4.26a)$$

and

Case 2 :

$$\varphi_\lambda(o,\infty) = \arg \gamma_\lambda - \frac{\mu}{v} . \arg(k) + \pi , \quad =1,2, \ldots v. \qquad (4.26b)$$

where

$\beta = \mu/v$, b, g and γ are taken from a straight line of segment V and from the related supporting polynomial.

13. Angles of arrival at a sink $s_j^q \neq o$ for $|k| = \infty$.
Rule 12 is applied to segment V of the transformed

equation $P(s-s_j^q;k) = 0$ (compare Rule 11).

14. Asymptotic behaviour for large $|k|$.

In line to Rule 1o we have :

Case 1 :

$$\varphi_\lambda(\infty,\infty) = \frac{1}{\nu}\cdot\left\{\arg\left(\frac{b}{g}\right) + \mu.\arg(k) + (2\lambda-1).\pi\right\} \qquad (4.27a)$$

for $\lambda = 1, \ldots, \nu$,

and

Case 2 :

$$\varphi_\lambda(\infty,\infty) = +\arg(\gamma_\lambda) + \frac{\mu}{\nu}\cdot\arg(k) , \qquad (4.27b)$$

where the constants b, g, μ, ν, β and γ_λ are taken from segment VIII of (4.2) and from the supporting polynomial related to this segment.

15. Asymptote points s_{ol} :

Case A : There are no points of the Exponent Diagram inside the area limited by the segment line and by the perturbation line. Then we have

$$s_{ol} = - \frac{\sum\limits_{j\in J} a^j_{n_j} \cdot \gamma_l^{n_j}}{\psi(\gamma_l)} , \text{ where}$$

the coefficients $a^j_{n_j}$ correspond to points on the perturbation line, γ_l is a non vanishing root of the supporting polynomial $\phi(\gamma_l)$ of the straight line considered (segment VIII), and $\psi(\gamma_l)$ is the derivative of $\phi(\gamma_l)$ with respect to γ at γ_l .

Case B : If condition A is hurt there exist no finite asymptote points s_{ol} .

16. Asymptotic behaviour for $k = 0$.

In line to Rule 12 we have :

Case 1 :

$$\varphi_\lambda(\infty, 0) = \frac{1}{\nu}.\{(\arg b - \arg g - \mu.\arg k - (2.\lambda - 1).\pi)\} + \pi \quad (4.28a)$$

and

Case 2 :

$$\varphi_\lambda(\infty, 0) = \arg \gamma_\lambda - \frac{\mu}{\nu}.\arg k + \pi \ , \ \lambda = 1, 2, \ldots, \nu \ , \quad (4.28b)$$

where the constants μ, ν, b, g and γ_λ are taken from the straight line of segment IV and the supporting polynomial related to this segment.

Note :

The asymptote points s_{ol} corresponding to Rule 16 are determined from the perturbation line and from the segment line of segment IV in analogy to Rule 15.

17. Angles of departure from s=o for finite real values $k_{lo}^j \neq 0$.

For each real value $k_{lo}^j \neq 0$ of $R_{lo}(k)$, Rule 1o is applied to the segment I of the transformed polynomial $P(s, k - k_{lo}^j)$.

18. Asymptote angles for non zero finite real roots k_n^j of $R_n(k)$

For each real value $k_n^j \neq 0$ of $R_n(k)$ Rule 14 is applied to segment VIII of the transformed polynomial $P(s, k - k_n^j)$.

Note :

The asymptote points s_{ol} corresponding to Rule 18 are computed in line to the note of Rule 16.

19. Breakaway or reentry points

They are among the common roots of the polynomials P(s,k) and $\partial P(s,k)/\partial s$ and may be computed as roots of the resultant of these two polynomials (compare (4.41)).

2o. <u>Return points</u>

The return points of root locus branches on the real axis are among the common roots of the polynomials P(s,k) and $\partial P(s,k)/\partial k$. They may be computed as roots of the resultant of these two polynomials (compare (4.42) and (4.43).

There exists an additional sequence of construction rules for higher order root locus plots which may be useful in specific problems but which are of minor value in standard situations of practice. (compare e.g. Steps 21 and 22). They are omitted here.

References

[4.1] W.R. Evans, Control System Synthesis by Root Locus Method, Trans. AIEE, 69, 195o.

[4.2] J.H. Blakelock, Automatic Control of Aircraft and Missiles, J.Wiley, 1965.

[4.3] G.Rosenau, Höhere Wurzelortskurven bei Mehrgrößen-Regelsystemen, IFAC Symposium, Düsseldorf, 1968.

[4.4] H.Schwarz, Optimale Regelung linearer Systeme, BI-Wissenschaftsverlag, 1976.

[4.5] H.Hahn, Zur Theorie und Technik singulärer Regelkreise, Habilitationsschrift, Universität Tübingen, Fachbereich Physik, 1977/78, Chapter VII.

[4.6] H.Hahn, compare [4.5] , Chapter XV.

[4.7] H.Hahn, compare [4.5], Chapter XIV.

[4.8] P.G. Retallack, Extended root locus technique for design of linear multivariable feedback systems, Proc. IEE, Vol.117, No.3, 197o.

[4.9] J.J.Beletrutti, A.G.J. McFarelane, Characteristic loci technique in multivariable control system design, Proc. IEE, Vol.118, No. 9, 1971.

[4.1o] H.Hahn, The Application of Root-Locus-Technique to Nonlinear Control Systems with Multiple Steady-States, Intern. J. of Control, 1977.

[4.11] H. Hahn, Comment and Corrections to [4.1o] , Intern. J. of Control, 1977.

4.12 B.Kouvaritakis, U. Shaked, Asymptotic behaviour of root loci of linear multivariable systems, Intern. J. Contr., 1976, Vol.23, No.3, pp. 297-34o.

[4.13] A.Tychonoff, Systeme von Differentialgleichungen, die bei den Ableitungen einen kleinen Parameter enthalten, Fortschritte der mathematischen Wissenschaften, Bd.VII, 1, (47), 1952, pp. 14o-142.

[4.14] F.Hoppensteadt, Singular Perturbations on the infinite interval, Tr. Am. Math.Soc. 123, 1966, pp. 521-534.

[4.15] H.Hahn, W.Ebinger, Singuläre Störungsrechnung kritischer Systeme, Preprint,Universität Tübingen, Fachbereich Physik, 1977.

[4.16] H. Hahn, compare [4.5] , Chapter IX.

[4.17] S. Butterworth, On the theory of filter amplifiers, Wireless Engr., 7, 193o, pp. 536-541.

V. Application of the Higher Order Root Locus Techniques to Simple Examples.

The higher order root locus construction rules derived in the preceding Chapter IV will now be applied to various simple examples (compare [5.1]).

For tutorial reasons it is desirable to treat examples each of which provides most of the phenomena of the root locus plots dicussed in Chapter IV. In general, only rather complex practical examples fulfil these requirements . As those examples are too lengthy to be treated within the framework of this monograph, the first three examples discuss simple formal polynomials of the form (4.2) which may be interpreted as characteristic polynomials of reduced linear multiloop control systems.

The residual examples discussed below, describe simplified versions of various practical problems from the fields of airplane dynamics, electrical network theory, control theory and vehicle dynamics.

In Chapter VI some of these examples are treated in more detail from the synthesis point of view.

1. Formal Examples

Example 5.1 (Formal characteristic polynomial)

Given a characteristic polynomial of the form

$$P(s,k) = s^3 + 3.s^2 + 2.s + k.s + k.b + k^2.a \quad .$$

Using some of the various higher order root locus construction rules of Chapter IV, this characteristic equation may be analysed as follows.

Rule 1 : Polynomials $Q_j(s)$ and $R_i(k)$; $j=0,1, \ldots ,2$; $i=0,1 \ldots ,3$.

$$Q_0(s) = s^3 + 3.s^2 + 2.s$$

$$Q_1(s) = s+b$$

$$Q_2(s) = a$$

$$R_0(k) = k.(k.a+b)$$

$$R_1(k) = 2+k$$

$$R_2(k) = 3, \; R_3(k) = 1 \quad .$$

It is useful, to write down the coefficients of the characteristic polynomial in form of table a. This table shows already implicitly the form of the exponent diagram of rule 3 and provides a connection between the various segments of the exponent polygon and the coefficients of the related supporting polynomials.

	s^0	s^1	s^2	s^3	
Q_2	a				k^2
Q_1	b	1			k^1
Q_0	o	2	3	1	k^0
	R_0	R_1	R_2	R_3	

Table 5.a.1: Coefficients of Polynomials $Q_j(s)$ and $R_i(k)$.

Rule 2 : Roots of the polynomials $Q_j(s)$ and $R_i(k)$.

$$Q_o(s) = s.(s+1).(s+2)$$
$$Q_1(s) = (s+b)$$

$$R_o(k) = a.k.(k+b/a)$$
$$R_1(k) = (k+2)$$

Again, it is useful to collect the roots of the above polynomials in tables b and c.

Q_i	i	$s-s_1^i$	$s-s_2^i$	$s-s_3^i$	
Q_1	1	(s+b)			k^1
Q_o	o	(s+o)	(s+1)	(s+2)	k^o
		R_o	R_1	R_2	

Table 5.b.1: Roots of polynomials $Q_j(s)$

	s^o	s^1	
Q_1	$k+{}^b/a$		$k-k_j^1$
Q_o	k+o	k+2	$k-k_j^o$
j	o	1	
R_j	R_o	R_1	

Table 5.c1: Roots of polynomials $R_i(k)$

Rules 3 and 4: Exponent Diagram

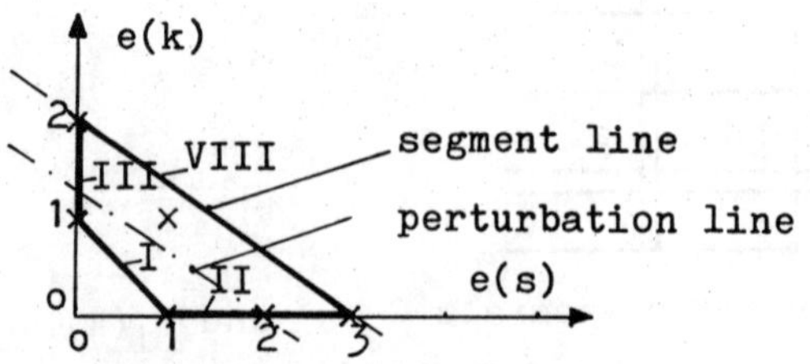

Figure 5.1.a: Exponent Diagram of $P(s;k) = o$

Rule 5 : Constants

$$l_o = 0 \quad , \quad n_o = n = 3 \quad , \quad n_q = q = 2 \quad ,$$

$$e_I = 1, \; e_{II} = 2, \; e_{VIII} = 3, \; \bar{e}_{III} = 1 \; , \quad .$$

Rule 7 : Slopes

$$\beta_I = \frac{\mu}{\nu} = 1/1 = 1 \quad , \quad \beta_{VIII} = \mu/\nu = 2/3$$

Note :

In agreement with the above stated rules, there exist $n=n_8=3$ root locus branches starting at roots $s_1^o=0$, $s_2^o=-1$ and $s_3^o=-2$ of $Q_o(s)$. As there does not exist a sink (zero) s_j^3 of the system, all branches tend to infinity for increasing values of k. An asymptote point s_{ol} does not exist.

Rule 1o : Angles of departure from $s_1^o = 0$ (segment I, k=o)

Assume $b \neq 0$, $b \neq -1$, $b \neq -2$ and $a \neq 0$. Then the polynomials $Q_o(s)$ and $Q_1(s)$ have no commom roots. Let $b = -3$ and $a = 2$. Then using rule 7($\mu = 1$, $\nu = 1$, $\beta_I = 1$) and equation (4.25b) we have

$$\varphi(o,o) = \arg\gamma + \beta.\arg k \qquad \text{or}$$

$$\varphi(o,o) = \arg(-b/2) + \arg k = \begin{cases} 0 & \text{for} \quad b < o \;, \quad k > o \\ \pi & \text{for} \quad b > o \;, \quad k > o \end{cases} ,$$

where $\gamma = -b/2 = 3/2 = 1{,}5$ is a root of the supporting polynomial $b + 2.\gamma = 0$ of segment I (compare Table 5.a.1).

Rule 11 : Angles of departure from $s_2^o = -1$ and $s_3^o = -2$ (segment II; k = o).

Replacing the variable s of $Q_o(s)$ and $Q_1(s)$ by the variables $s' + s_2^o = s' - 1$ and $s' + s_3^o = s' - 2$ yields for

$\underline{s_2^0 = -1:}$

$$Q_0'(s') = 2.(s'-1)+3.(s'-1)^2+(s'-1)^3=0-1.s'+ \ldots ,$$

$$Q_1'(s') = b+s'-1=(b-1)+s' \quad ,$$

$$\gamma = -(b-1)/(-1)=b-1 = -4 \quad , \quad \beta_I = 1/1 = 1 \quad \text{and}$$

$$\varphi(-1,0) = \arg\gamma + \beta.\arg k = \begin{cases} 0 & \text{for} \quad b>1 \ , \ k>0 \\ \pi & \text{for} \quad b<1 \ , \ k>0 \end{cases} ,$$

and for

$\underline{s_3^0 = -2:}$

$$Q_0'(s') = 2.(s'-2)+3.(s'-2)^2+(s'-2)^3=0+2.s'+ \ldots ,$$

$$Q_1'(s') = (b-2)+s' \quad ,$$

$$\gamma = -(b-2)/2 = 1-b/2 = 5/2 \quad , \quad \beta_I = 1/1 = 1 \quad \text{and}$$

$$\varphi(-2,0) = \arg\gamma + \beta.\arg k = \begin{cases} 0 & \text{for} \quad b<2 \ , \ k>0 \\ \pi & \text{for} \quad b>2 \ , \ k>0 \end{cases}$$

<u>Rule 14 :</u> <u>Asymptotic behaviour for large values of $|k|$.</u>

In line with rule 7 ($\beta_{VIII} = 2/3$), equation (4.27b) yields

$$\varphi(\infty,\infty) = \arg(\gamma) + \beta_{VIII}.\arg k$$

or using equation (4.27b) and the supporting polynomial of segment VIII

$$\gamma^3 + a = 0 \quad \text{(compare Table 5.a.1)} :$$

$$\varphi_\lambda(\infty,\infty) = \arg\gamma_\lambda + \beta_{VIII}.\arg k \quad , \quad \lambda = 1,2,3$$

or

$$\varphi_1(\infty,\infty) = \frac{\pi}{3}$$

$$\varphi_2(\infty,\infty) = \pi \qquad \text{for} \quad a>0 \quad , \quad k>0$$

$$\varphi_3(\infty,\infty) = \frac{5}{3}\cdot\pi$$

and

$$\varphi_1(\infty,\infty) = 0$$

$$\varphi_2(\infty,\infty) = \frac{2}{3}\cdot\pi \qquad \text{for} \quad a<o \quad , \quad k>o$$

$$\varphi_3(\infty,\infty) = \frac{4}{3}\cdot\pi$$

Rule 15 : Asymptote points

There is a point of the exponent diagram inside the area, limited by the segment line and the perturbation line. As a consequence, finite asymptote points s_{o1} do not exist.

Rule 17 : Departure of root locus branches from $s_1^o=0$ for roots $k_{1_o}^j$ of $R_{1_o}(k)$ (segment III, $k_{1_o}^j \neq 0$).

Table 5.1.c shows that $k_o^1=0$ and $k_o^2=-b/a$ are roots of the polynomial $R_o(k)$. Therefore a root locus branch meets the point $s=0$ for parameter values $k_o^1=0$ and $k_o^2=-b/a=1,5$.

Note :

Rule 17 in connection with rules 1o and 11 already show that for $(a,b)=(2,-3)$ there exists at least one return point on the real axis.

Rule 2o : Return points

From equations (4.42b) and (4.43) we have for $q=2$:

$$R(P\,,\ \partial P/\partial k\ ) = Q_2.(Q_1^{\,2}-4\ Q_0\ .\ Q_2)$$

$$R = a.\ ((s+b)^2-\ 4.a.s.(s+1).(s+2)\ \ )$$

$$R = a.\ (-4.a.s^3 + (1-12.a).s^2 +(2.b-8.a).s+b^2\ )\ ,$$

or using $(a,b) = (2,-3)$,

$$R = 2.(\ -8.s^3-23.s^2-22.s + 9\ )\quad .$$

For positive values of k the point $s_1 = 0,303$ is the only real return point. The points

$s_2 = -\ 1,59 + i.1,09$ and $s_3 = -\ 1,59 - i.1,09$ are complex conjugate return points for $k > 0$. A rough sketch of the root locus plot based on the above rules is drawn in Figure 5.1b for $k \geq 0$. The exact root locus plot is shown in Figure 5.1c for $k \geq 0$ (compare example 6.1 of Chapter VI for further discussions).

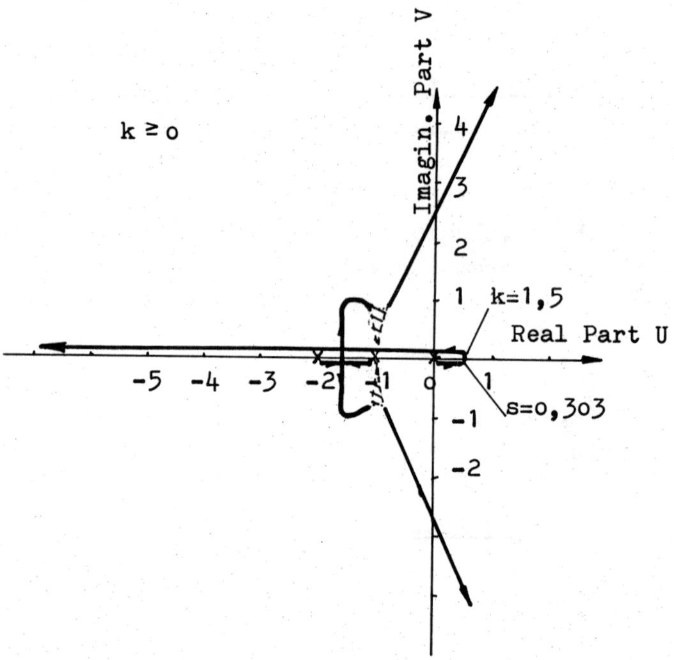

Figure 5.1.b: Qualitative root locus plot (a,b) = (2,-3).

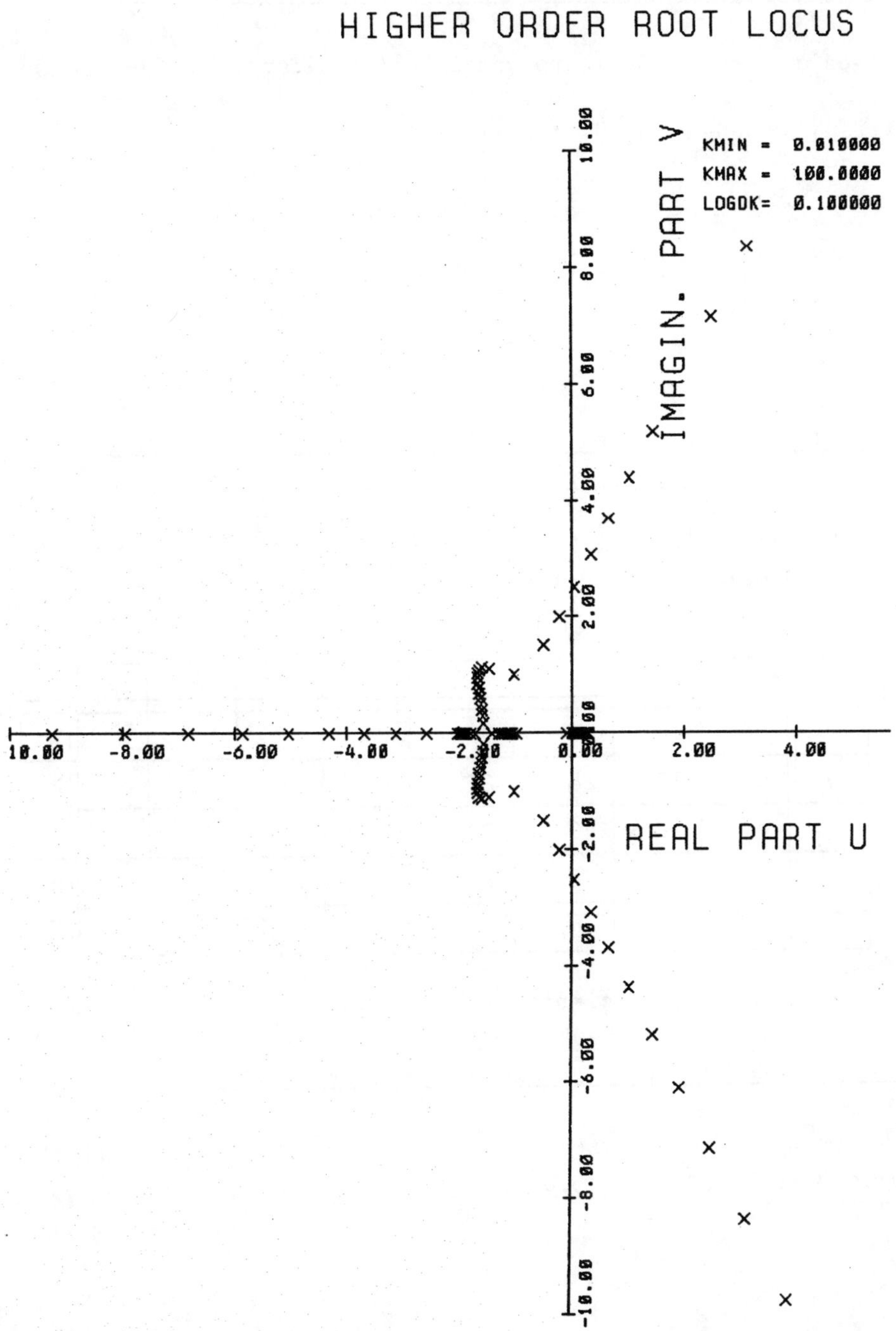

Figure 5.1.c: Computer root locus plot (a,b)=(2,-3).

Example 5.2 (Formal characteristic polynomial)

Given a characteristic polynomial of the form

$$P(s;k) = 20.s^3+ 29.s^4+ 10.s^5+ s^6$$

$$+ (72.s+246.s^2+329.s^3+220.s^4+78.s^5+14.s^6+s^7).k$$

$$+ (12.s+28.s^2+23.s^3+8.s^4+s^5).k^2$$

$$+ (s^2+s^3).k^3$$

Rule 1 : Polynomials $Q_j(s)$ and $R_i(k)$; j=o, 1, ... ,3; i=o, ... ,7 .

The coefficients of these polynomials are collected in Table 5.a.2 (comp. Example 5.1).

	s^0	s^1	s^2	s^3	s^4	s^5	s^6	s^7	
Q_3	o	o	1	1					k^3
Q_2	o	12	28	23	8	1			k^2
Q_1	o	72	246	329	22o	78	14	1	k^1
Q_0	o	o	o	2o	29	1o	1	o	k^0
	R_0	R_1	R_2	R_3	R_4	R_5	R_6	R_7	

Table 5.a.2 : Coefficients of Polynomials $Q_j(s)$ and $R_i(k)$.

Rule 2 : Roots of polynomials $Q_j(s)$ and $R_i(k)$.

The roots of the polynomials $Q_i(s)$ and $R_j(k)$ are collected in Tables 5.b.2 and 5.c.2.

Q_i	i	$s-s_1^i$	$s-s_2^i$	$s-s_3^i$	$s-s_4^i$	$s-s_5^i$	$s-s_6^i$	$s-s_7^i$	
Q_3	3	s+o	s+o	s+1					k^3
Q_2	2	s+o	s+1	s+2	s+2	s+3			k^2
Q_1	1	s+o	s+1	s+1	s+2	s+3	s+3	s+4	k^1
Q_0	o	s+o	s+o	s+o	s+1	s+4	s+5		k^0
		R_0	R_1	R_2	R_3	R_4	R_5	R_6	

Table 5.b.2 : Roots of polynomials $Q_j(s)$

	s^0	s^1	s^2	s^3	s^4	s^5	s^6	s^7	
Q_2			k+14+ -i.7,o7	k+11,5 -i.14					$k-k_j^2$
Q_1		k+6	k+14+ +i.7,o7	k+11,5 +i.14	k+27,4	k+77,9			$k-k_j^1$
Q_0		k+o	k+o	k+o,o6	k+o,13	k+o,128	k+o,o7	k+o	$k-k_j^0$
j	o	1	2	3	4	5	6	7	
R_j	R_0	R_1	R_2	R_3	R_4	R_5	R_6	R_7	

Table 5.c.2: Roots of polynomials $R_i(k)$

Rules 3 and 4 : Exponent Diagram

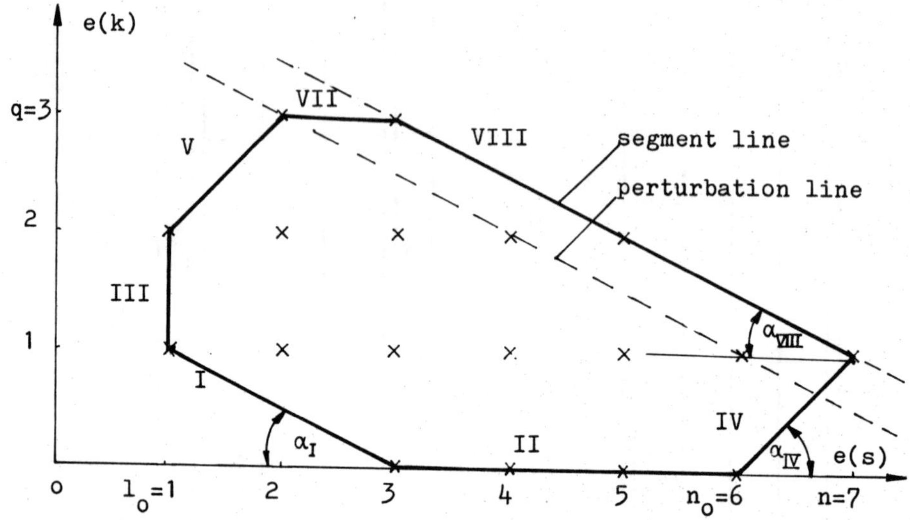

Fig. 5.2.a: Exponent Diagram of $P(s;k) = o$.

Rule 5 : Constants

$l_o = 1$, $n_o = 6$, $n = 7$, $q = 3$

$e_I = 2$, $e_{II} = 3$, $e_{IV} = 1$, $e_V = 1$, $e_{VII} = 1$, $e_{VIII} = 4$

$\bar{e}_{III} = 1$, $\bar{e}_{VI} = o$.

Rule 7 : Slopes

$\beta_I = 1/2 = 0,5$; $\beta_{IV} = 1/1 = 1$; $\beta_V = 1/1 = 1$; $\beta_{VIII} = 2/4 = 0,5$.

Note :

In agreement with the above stated rules, there exist $n = 7$ root locus branches, a complete pole sink compensation at $s = o$ and at $s = -1$ and partial pole sink compensations at $s = -2$, $s = -3$ and $s = -4$. The points $s = o$ and $s = -1$ are degenerated root locus branches. There exist

two finite sinks at $s = 0$ and at $s = -1$ and four asymptotes which meet finite asymptote points $s_{ol}(l=1,2,3,4)$.
There are root locus branches which meet the point $s = 0$ for $k = -6$ and for $k = 0$. One root locus branch starts at infinity for $k = 0$. The other branches start at the finite points s_j^o (roots of $Q_o(s)$).

<u>Rule 1o :</u> <u>Angles of departure from $s_j^o = 0$ (segment I, k=o).</u>

Exactly two root locus branches start from point $s = 0$. The angles of departure from this point are computed from (4.25b), using the supporting polynomial of segment I

$$2o.\gamma^2 + 72 = 0 \quad \text{with roots} \quad \gamma_{1/2} = \pm\, i.\sqrt{3,6}\,:$$

$$\varphi_1(o,o) = \arg \gamma_1 + \beta_I.\arg k = \pi/2 \quad , \quad \text{for} \quad k > o$$

and

$$\varphi_2(o,o) = \arg \gamma_2 + \beta_I.\arg k = -\pi/2 \quad , \quad \text{for} \quad k > o \quad .$$

<u>Rule 11 :</u> <u>Angles of departure from $s_5^o = -4$ and $s_6^o = -5$ (segment II, k = o)</u>

Replacing the variable s of $Q_o(s)$ and $Q_1(s)$ by the variables $s'-4$ and $s'-5$ yields for

<u>$s_5^o = -4$:</u>

$$Q_o'(s') = o + 192s' + \ldots$$

$$Q_1'(s') = o + \ldots \quad ,$$

$$Q_2'(s') = -460 + \ldots \quad ,$$

$$\gamma = 460/192 \quad , \quad \beta_I = 2/1 = 2 \quad \text{and}$$

$$\varphi(-4,o) = \arg\gamma + \beta.\arg k = 0 \quad \text{for} \quad k > o \quad ,$$

and for

$s_6^o = -5$:

$Q_o'(s') = o - 2ooo.s' \ldots$,

$Q_1'(s) = -91o + \ldots$,

$\gamma = -91o/2ooo < o$, $\beta_I = 1/1 = 1$ and

$\varphi(-5,o) = \arg\gamma + \beta.\arg k = \pi$ for $k > o$.

Note :

The roots $s_3^o = o$ and $s_4^o = -1$ of $Q_o(s)$ are degenerated root locus branches. There exist no angles of departure of branches from these points for $k = o$.

Rule 12 : Angles of arrival at $s = o$ (segment V , $k = \infty$).

For $k = \infty$ one root locus branch tends towards the point s=o. The related angle of arrival is computed according to equation (4.26b)

$$\varphi(o,\infty) = \arg\gamma + \beta_V.\arg k + \pi \quad ,$$

where γ is solution of the polynomial $\gamma + 12 = o$.

Then the angle of arrival at s=o for $k=\infty$ is computed as

$$\varphi(o,\infty) = o \quad \text{for} \quad k > o.$$

Rule 14 : Asymptote angles (segment VIII, $k = \infty$)

Using $\beta_{VIII} = 1/2$, the supporting polynomial of segmentVIII

$$\gamma^4 + \gamma^2 + 1 = o \quad ,$$

with roots

$$\gamma_1 = o,5 + i.o,87$$
$$\gamma_2 = o,5 - i.o,87$$
$$\gamma_3 = -o,5 + i.o,87$$
$$\gamma_4 = -o,5 - i.o,87$$

and relation (4.27b) yields the asymptote angles

$$\varphi_1(\infty,\infty) = +\ 60^o$$

$$\varphi_2(\infty,\infty) = -\ 60^o$$

$$\varphi_3(\infty,\infty) = +\ 120^o$$

$$\varphi_4(\infty,\infty) = -\ 120^o \quad .$$

<u>Rule 15 :</u> <u>Asymptote points s_{ol}</u>

Using relation (4.37a) in connection with Figure 5.2a and the roots γ_λ ($\lambda = 1,2,3,4$) of the supporting polynomial related to segment VIII yields the asymptote points

$$s_{o\lambda} = \left. \frac{14.\gamma^6 + 8.\gamma^4 + \gamma^2}{7.\gamma^6 + 5.\gamma^4 + 3.\gamma^2} \right|_{\gamma = \gamma_\lambda}$$

or

$$s_{o1} = s_{o3} = -\ 3{,}22 + i.0{,}16$$

$$s_{o2} = s_{o4} = -\ 3{,}22 - i.0{,}16$$

<u>Rule 16 :</u> <u>Asymptote angle for $k = o$ (segment IV).</u>

Using the relations $\gamma = -1$ and $\beta_{IV} = 1$ (compare Table 5.a.2 and Figure 5.2.a) equation (4.28b) yields :

$$\varphi(\infty,0) = \arg\gamma + \beta_{IV}.\arg k + \pi = o \quad \text{for} \quad k > o \quad .$$

<u>Rule 17 :</u> <u>Departure of root locus branches from $s_j^o = o$</u>
<u>for roots k_{lo}^j of R_{lo} (segment III; $k_{lo}^j \neq o$).</u>

As is shown in Table 5.c.2, the point $s = o$ is met by root locus branches for $k = o$ and for $k = -\ 6.$ (comp. roots of $R_{lo}(k)$, $l_o = 1$).

Rule 2o : Return points

Equations (4.42b) and (4.43) imply for q = 3 that besides of the sinks of the system (s_1^3 = o and s_2^3 = - 1) the point s = - 4 (as a common root of $Q_0(s)$ and $Q_1(s)$) is a real return point of root locus branches. Further return points may be computed from equation (4.43).
A rough sketch of the root locus plot based on the above rules, and an exact plot are drawn in Figures 5.2b and 5.2c, respectively for k ≥ o).

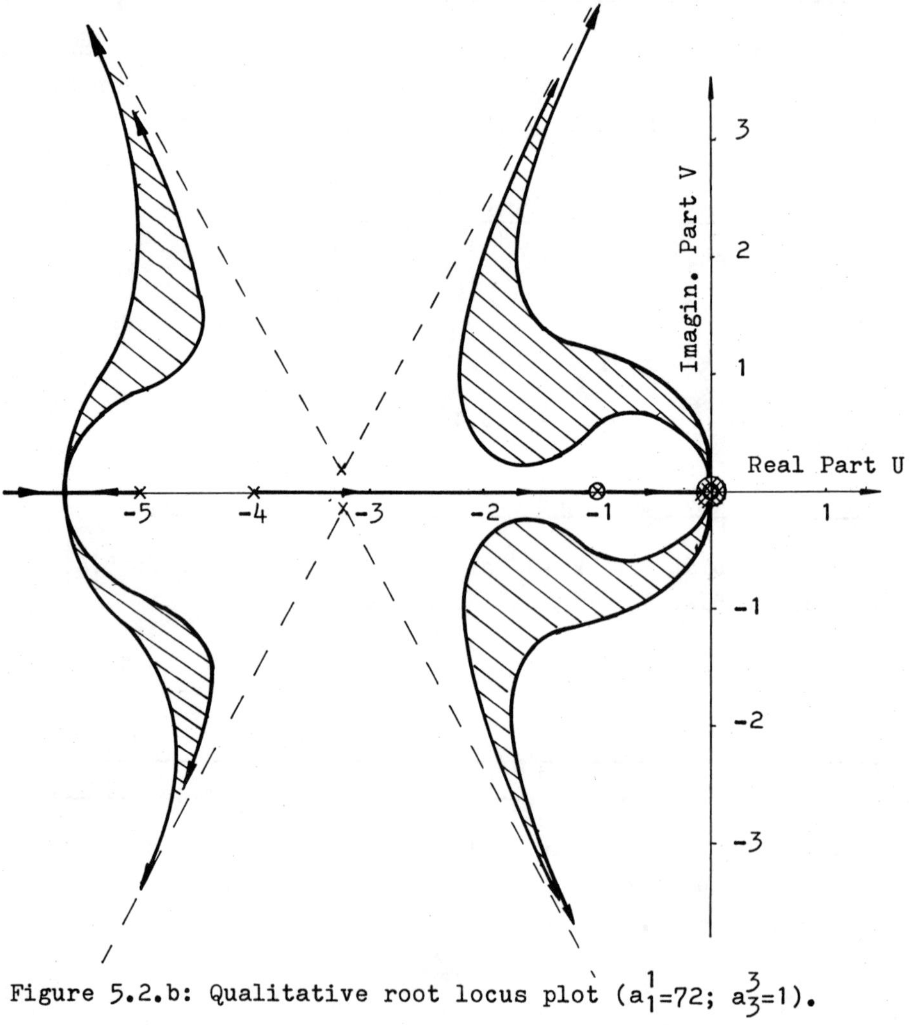

Figure 5.2.b: Qualitative root locus plot (a_1^1=72; a_3^3=1).

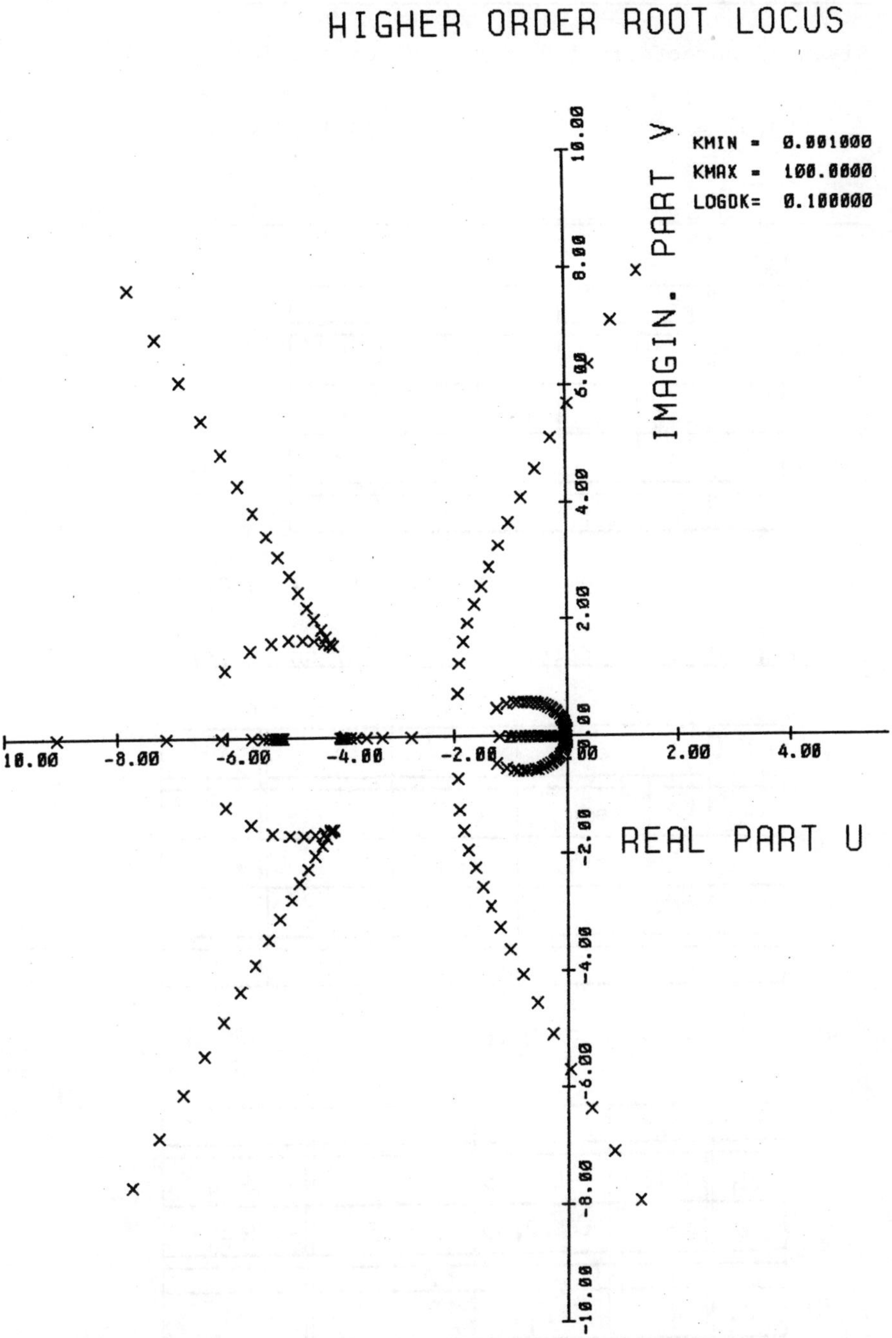

Figure 5.2.c: Computer root locus plot(a_1^1=72, a_3^3=1).

Example 5.3 (Formal characteristic polynomial)

Given a characteristic polynomial of the form

$$P(s,k) = s^3+3.s^2+2.s-16.k+k.s^2+k^2+0,25.k^2.s$$

Rule 1 : Polynomial coefficients of $Q_j(s)$ and $R_i(k)$

	s^0	s^1	s^2	s^3	
Q_2	1	0,25			k^2
Q_1	-16	0	1		k^1
Q_0	0	2	3	1	k^0
	R_0	R_1	R_2	R_3	

Table 5.a.3 : Polynomial coefficients of $Q_j(s)$ and $R_i(k)$

Rule 2 : Roots of polynomials $Q_j(s)$ and $R_i(k)$

Q_i	i	$s-s_1^i$	$s-s_2^i$	$s-s_3^i$	
Q_2	2	s+4			k^2
Q_1	1	s+4	s-4		k^1
Q_0	0	s+0	s+1	s+2	k^0
		R_0	R_1	R_2	

Table 5.b.3 : Roots of $Q_j(s)$

	s^0	s^1	s^2	
Q_1	k-16	k-i.2,85		$k-k_j^1$
Q_0	k+0	k+i.2,85	k+0,33	$k-k_j^0$
j	0	1	2	
R_j	R_0	R_1	R_2	

Table 5.c.3 : Roots of $R_i(s)$

Rules 3 and 4 : Exponent Diagram

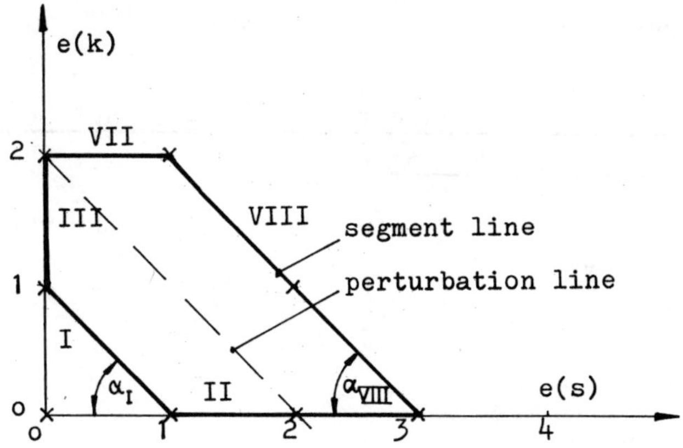

Figure 5.3.a : Exponent Diagram of $P(s;k)=0$

Rule 5 : Constants

$l_o = 0$, $n_o = n = 3$, $q = 2$

$e_I = 1$, $e_{II} = 2$, $e_{VIII} = 2$, $e_{VII} = 1$, $\bar{e}_{III} = 1$

Rule 7 : Slopes

$\beta_I = - - = \frac{1}{1} = 1$, $\beta_{VIII} = \frac{\mu}{\nu} = \frac{2}{2} = 1$

Note :

There exist $n = n_o = 3$ root locus branches starting at the poles $s_1^o = 0$, $s_2^o = -1$ and $s_3^o = -2$, two of which tend to infinity, and one tending towards the sink $s_1^2 = -4$. For $k_o^1 = 0$ and $k_o^2 = 16$ a root locus branch meets the point $s_1^o = 0$. There exist two finite asymptote points.

Rule 1o : Angles of departure from $s_1^o = 0$ (segment I, $k = 0$)

One root locus branch starts from the point $s_1^o = 0$. Its angle of departure follows from equation (4.25b) in connection with the constants

$$\beta_I = 1 \quad \text{and} \quad \gamma = 8$$

corresponding to segment I. We have

$$\varphi(o,o) = \arg\gamma + \beta.\arg k = o \quad \text{for} \quad k > o$$

<u>Rule 11 :</u> <u>Angles of departure from $s_2^o = -1$ and $s_3^o = -2$ (segment II, k = o)</u>

Replacing the variable s of $Q_o(s)$ and $Q_1(s)$ by the variables s'- 1 and s'- 2 yields for :

<u>$s_2^o = -1$:</u>

$$Q_o(s') = o - 1.s' + \ldots \quad ,$$

$$Q_1(s') = -17 + s' \quad ,$$

$$\gamma = -17 \; , \; \beta = 1 \quad \text{and} \quad \text{(compare 4.25b)} \quad ,$$

$$\varphi(-1,o) = \pi \quad \text{for} \quad k > o$$

and for

<u>$s_3^o = -2$:</u>

$$Q_o(s') = o + 2.s' + \ldots$$

$$Q_1(s-) = -18 + s'$$

$$\gamma = 9 \quad , \quad \beta = 1 \quad \text{and}$$

$$\varphi(-2,o) = o \quad \text{for} \quad k > o \quad .$$

<u>Rule 13 :</u> <u>Angles of arrival at $s_1^2 = -4$ (segment VII, k $=\infty$).</u>

Replacing the variable s of $Q_2(s)$ and $Q_o(s)$ by s'- 4 yields the relations ($Q_1(s)$ is omitted as $Q_1(s)$ and $Q_2(s)$ have a common root s = - 4):

$Q_2'(s) = 0 + 0{,}25s'$,

$Q_0'(s) = -\ 24 + \ldots$,

$\gamma = \frac{24}{0{,}25} = 96 > 0$, $\beta_V = 2$

and (compare 4.26b)

$\varphi(-\ 4, \infty) = \pi$ (for orbitrary values of k).

Rule 14 : Asymptote angles (segment VIII, $k = \infty$)

Using $\beta_{VIII} = 1$, $\gamma^2 + \gamma + 1/4 = 0$ or

$\gamma_1 = \gamma_2 = -\ 1/2$ yields in connection with (4.27b)

$\varphi_1(\infty, \infty) = \varphi_2(\infty, \infty) = \pi$ for $k > 0$.

Rule 15 : Asymptote points s_{oj}

Because of the results of Rule 14, the asymptote points are not needed.

Rule 17 : Departure of root locus branches from $s_j^o = 0$ (segment III, $k = k_{1_o}^j$)

For k = o and k = 16 a branch meets point s = o (compare Table 5.c.3).

Rule 2o : Return points

Using the relation (compare 4.43)

$R(P, \partial P/\partial k) = Q_2 \cdot (Q_1^2 - 4 . Q_0 \cdot Q_2)$

or

$R = 0{,}25.(s+4).((s+4)^2.(s-4)^2 - 4.s.(s+1).(s+2).0{,}25.(s+4))$

yields

$$R = 0{,}25.(s+4)^2.(\ (s+4).(s-4)^2-s(s+1).(s+2)\ )$$

$$= 0{,}25.(s+4)^2.(s^3 - 16s+64-s^3-3s^2-2s)$$

$$= 0{,}25.(s+4)^2.(-7s^2-18s+64)$$

$$= - 1{,}75.(s+4)^2.(s+4{,}56).(s-2) \quad .$$

The points $s = - 4$, $s = + 2$ and $s = - 4{,}56$ are return points of root locus branches ($s = - 4$ is a sink of the system).
The related root locus plots are drawn in Figures 5.3.b and 5.3.c for $k \geq 0$.

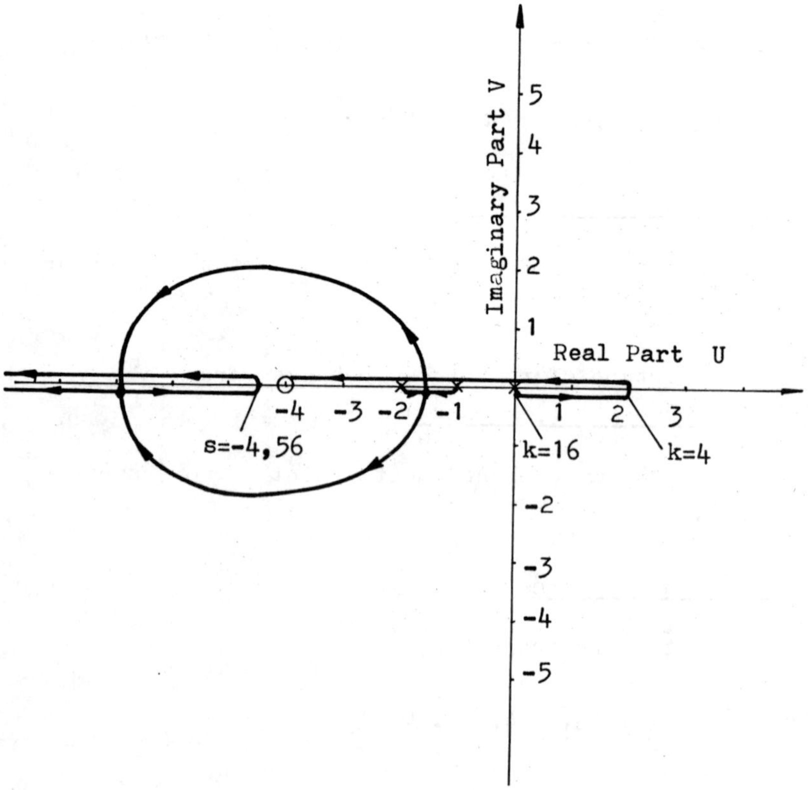

Figure 5.3.b: Qualitative root locus plot (a_1^2=o,25).

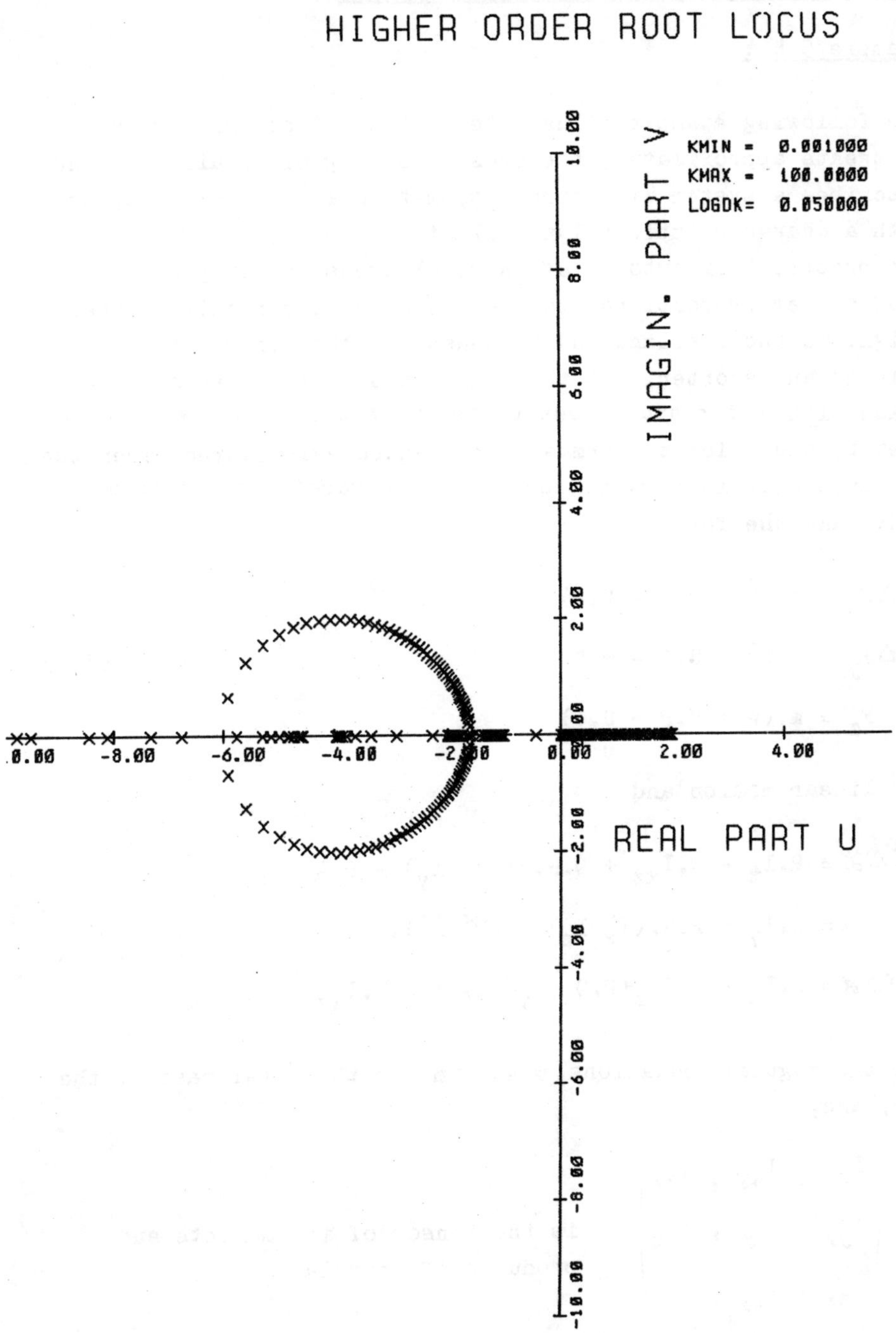

Figure 5.3.c: Computer root locus plot (a_1^2 = o,25).

2. Inertial cross coupling of an aircraft

Example 5.4 :

The following example illustrates point (i) of Chapter IV.1. It treats approximately the cross coupling of an aircraft and describes a system with strong symmetrical internal coupling with a characteristic polynomial of the form (4.2).
The practical importance of inertial cross coupling of aircrafts grew proportional to the tendency to concentrate their weight in the fuselage and to construct the aircraft's wings thinner and shorter. Let the axis fixed to the aircraft are taken with X forward, Y out the right wing, and Z downward as seen by the pilot to form a right handed axis system. Then the complete equations of motion for the aircraft (as a rigid body) take the form

$$\begin{aligned} \Sigma\Delta F_x &= m.(\dot{U} + W.Q - V.R) \\ \Sigma\Delta F_y &= m.(\dot{V} + U.R - W.P) \\ \Sigma\Delta F_z &= m.(\dot{W} + V.P - U.Q) \end{aligned} \tag{5.1}$$

for linear motion and

$$\begin{aligned} \Sigma\Delta\mathcal{L} &= \dot{P}.I_x - \dot{R}.I_{xz} + Q.R.(I_z - I_y) - P.Q.I_{xz} \\ \Sigma\Delta\mathcal{M} &= \dot{Q}.I_y + P.R.(I_x-I_z) + (P^2-R^2).I_{xz} \\ \Sigma\Delta\mathcal{N} &= \dot{R}.I_z - \dot{P}.I_{xz}+P.Q.(I_y-I_x) + Q.R.I_{xz} \end{aligned} \tag{5.2}$$

for the angular equations, where m is the total mass of the airplane,

$$\begin{pmatrix} I_x , & I_{xy} , & I_{xz} \\ I_{yx} , & I_y , & I_{yz} \\ I_{zx} , & I_{zy} , & I_z \end{pmatrix}$$ is the tensor of its moments and products of inertia

and (5.3)

$$\begin{pmatrix} \Delta F_x \\ \Delta F_y \\ \Delta F_z \end{pmatrix}, \begin{pmatrix} \Delta \mathcal{L} \\ \Delta \mathcal{M} \\ \Delta \mathcal{N} \end{pmatrix}, \begin{pmatrix} P \\ Q \\ R \end{pmatrix} \text{ and } \begin{pmatrix} U \\ V \\ W \end{pmatrix} \text{ are the}$$

vectors of the external forces, moments, angular velocities and linear velocities, respectively.

Equation (5.2) only holds on the assumption

$$- I_{yx} = I_{xy} = I_{yz} = 0 \quad . \tag{5.4}$$

A variety of dynamic phenomena of aircrafts may be described by linearizing and separating the six simultaneous non linear equations (5.1) and (5.2) (compare [5.2]).
On the other hand, there exist phenomena which can not be described by the linearized equations. Let for instance, I_x become much smaller than I_y and I_z . Then the terms (I_x-I_z) and (I_y-I_x) of (5.2) become large. If a rolling moment is introduced, this results in some yawing moment, and the term $P.R(I_x-I_z)$ may become large enough to cause an uncontrollable pitching moment. This inertial cross-couplings are well known in practice. Some aircrafts are completely unstable at high roll rates. A rigorous mathematical analysis of these phenomena has to be based on the nonlinear system equations (5.1) and (5.2).
Approximate results of these coupling phenomena which are in excellent agreement with experiments may be achieved by making certain simplifying assumptions. As it is desired to study the effects of high-roll rates, $P = P_o$ is assumed to become a parameter k. This assumption is justified by the fact that zero roll rate motions represent the short-period modes whereas high-roll rates introduce additional long-period modes.

On the assumptions

$$U = U_o = \text{constant} \quad ,$$

$$v := V - V_o \quad , \quad w := W - W_o \quad ,$$

$$p := P - P_o \quad , \quad q := Q = Q_o \quad , \quad r := R = R_o \quad , \qquad (5.5)$$

$$V_o = W_o = Q_o = R_o = 0 \quad ,$$

$$M = N = F_y = F_z = 0 \quad ,$$

and if the stability axes are considered as principle axes, then Equations (5.1) and (5.2) become

$$\begin{aligned} &\dot{q}.I_y + k.r.(I_x - I_z) - c_4.q - c_5.\dot{\alpha} - c_6.\alpha = 0 \\ &\dot{r}.I_z + k.q.(I_y - I_x) - c_1.r - c_2.\beta = c_3.k \end{aligned} \qquad (5.6)$$

$$\begin{aligned} &\dot{\beta} + r - k.\alpha = 0 \\ &\dot{\alpha} + k.\beta - q = o \quad , \end{aligned} \qquad (5.7)$$

where

$$\beta \cong v/U_o \quad \text{and} \quad \alpha \cong w/U_o \quad ,$$

and the terms in connection with the constants c_1 to c_6 represent the aerodynamic moments of the two modes.

Substitution (5.7) and its time derivative into (5.6), regrouping and taking the Laplace transformation yields :

$$\begin{aligned} &(a_1'.s^2 + a_2'.s + a_0'.k^2).\beta(s) + (c_1.s + c_0).k.\alpha(s) = o \\ &(a_3'.k.s + a_4'.k).\beta(s) + (c_3.s^2 + c_4.s + c_5.k^2).\alpha(s) = o \quad . \end{aligned} \qquad (5.8)$$

Using realistic data, the characteristic equation correspon-ding to (5.8) has the form (compare [5.2] , [5.3]):

$$P(s,k) = (a_4^0.s^4 + a_3^0.s^3 + a_2^0.s^2 + a_1^0.s + a_0^0)$$

$$+ k.(a_2^1.s^2 + a_1^1 s + a_0^1) + k^2.a_0^2 \quad ,$$

where

$$a_0^0 = 53{,}9811,\ a_1^0 = 4{,}184,\ a_2^0 = 14{,}7665,\ a_3^0 = 0{,}5843,\ a_4^0 = 1{,}0$$

$$a_0^1 = -\ 13{,}2416,\ a_1^1 = 0{,}3108,\ a_2^1 = 1{,}9032,\ a_0^2 = 0{,}817 \quad .$$

Now the root locus construction rules will be applied to P(s,k).

<u>Rule 1 : Polynomial coefficients of $Q_j(s)$ and $R_i(k)$.</u>

	s^0	s^1	s^2	s^3	s^4	
Q_2	o,817					k^2
Q_1	-13,25	o,311	1,9			k^1
Q_0	54,o	4,184	14,77	o,584	1,o	k^0
	R_0	R_1	R_2	R_3	R_4	

Table 5.a.4 : Polynomial coefficients of $Q_j(s)$ and $R_i(k)$.

<u>Rule 2 : Roots of polynomials $Q_j(s)$ and $R_i(k)$.</u>

Q_j	i	$s-s_1^i$	$s-s_2^i$	$s-s_3^i$	$s-s_4^i$	
Q_1	1	s-2,6	s+2,72			k^1
Q_0	o	s+o,23 +i.2,76	s+o,23 -i.2,76	s+o,o6 +i.2,65	s+o,o6 -i.2,65	k^0
		R_0	R_1	R_2	R_3	

Table 5.b.4 : Roots of $Q_j(s)$.

	s^0	s^1	s^2	
Q_1	k-8,1 +i.o,66			$k-k_j^1$
Q_0	k-8,1 -i.o,66	k+13.4	k+7.7	$k-k_j^0$
j	o	1	2	
R_j	R_0	R_1	R_2	

Table 5.c.4: Roots of $R_i(k)$

Rules 3 and 4 : Exponent Diagram

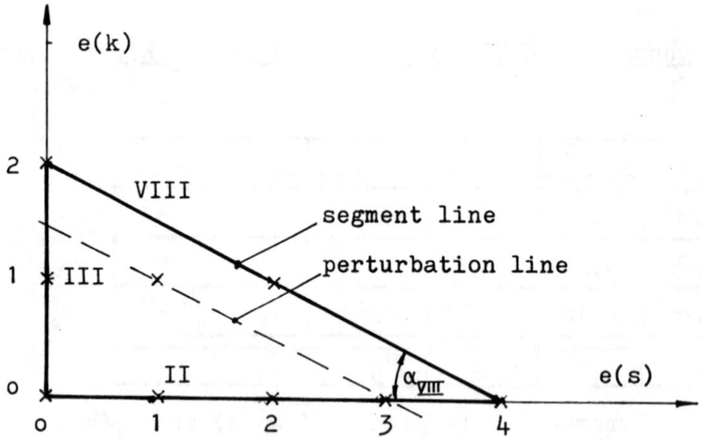

Figure 5.4.a : Exponent Diagram

Rule 5 : Constants

$$l_0 = 0\ ,\ q = 2\ ,\ n = 4\ ,\ n_0 = 4$$

$$e_I = 0\ ,\ e_{II} = 4\ ,\ e_{IV} = 0\ ,\ e_V = 0$$

$$e_{VII} = 0\ ,\ e_{VIII} = 4\ ,\ \bar{e}_{III} = 2\ ,\ \bar{e}_{VI} = 0\ .$$

Rule 7 : Slopes

$$\beta_{VIII} = \frac{\mu}{\nu} = \frac{1}{2} \quad .$$

Note :

There exist $n = n_0 = 4$ root locus branches starting at poles s_1^0 to s_4^0, all of which tend to infinity for increasing values of k (there does not exist a sink of the system).

Rule 11 : Angles of departure from s_1^0, s_2^0, s_3^0, and s_4^0 (segment II, k = o)

Replacing the variable s in $Q_0(s)$ and $Q_1(s)$ by $s' + s_i^0$, i = 1,2,3,4 yields for

$s_1^0 = - 0,23 + i.2,76$:

$$Q_0(s') = (5,2 - i.3,15).s' + \dots \quad ,$$

$$Q_1(s') = 1,9.(-14.54 - i.0,84 + \dots \quad)$$

and $\beta = 1$ (compare 4.25b)

$$\varphi(-0,23 + i.2,76 \; ; \; o) = 33^o \quad \text{for} \quad k > o \quad ,$$

for

$s_2^0 = - 0,23 - i.2,76$:

$$\varphi(-0,23 - i.2,76 \; ; \; o) = 2.\pi - 33^o \quad \text{for} \quad k > o \quad ,$$

for

$s_3^0 = - 0,06 + i.2,65$:

$$Q_0(s') = (4,76 - i.3,3).s' + \dots \quad ,$$

$$Q_1(s') = 1,9.(-13,9 + i.0,103 + \dots \;)$$

and $\beta = 1$

$$\varphi(-0{,}06 + i.2{,}65\ ;\ 0) = \pi + 33^o \quad \text{for} \quad k > 0$$

and for

<u>$s_3^o = -\ 0{,}06 - i.2{,}65$:</u>

$$\varphi(-0{,}06 - i.2{,}65\ ;\ 0) = \pi - 33^o \quad \text{for} \quad k > 0 \ .$$

<u>Rule 14 :</u> <u>Asymptote angles (segment VIII, $k = \infty$)</u>

The roots of the supporting polynomial of segment VIII

$$1.\gamma^4 + 1{,}9.\gamma^2 + 0{,}817 = 0$$

are

$$\gamma_1 = i.0{,}79 \ , \quad \gamma_2 = -\ i.0{,}79$$

$$\gamma_3 = i.1{,}12 \ , \quad \gamma_3 = -\ i.1{,}12 \ .$$

Then, using $\beta_{VIII} = 2/4 = 0{,}5$ and equation (4.27b) ,

the asymptote angles are

$$\left.\begin{aligned} \varphi_1(\infty,\infty) &= \pi/2 \\ \varphi_2(\infty,\infty) &= -\ \pi/2 \\ \varphi_3(\infty,\infty) &= \pi/2 \\ \varphi_4(\infty,\infty) &= -\ \pi/2 \end{aligned}\right\} \quad \text{for} \quad k > 0 \ .$$

<u>Rule 15 :</u> <u>Asymptote points s_{oj}</u>

Equation (4.37a) and Figure 5.4.a yield in connection with γ_λ from rule 14 :

$$s_{o_\lambda} = -\ \frac{0 + 0{,}584.\gamma_\lambda^{\ 3} + 0{,}31.\gamma_\lambda^{\ 1}}{4\ .\ 1.\gamma_\lambda^{\ 3} + 2.\ 1{,}90\ .\ \gamma_\lambda^{\ 1} + 0} \qquad \text{or}$$

$$s_{o\lambda} = - \frac{0,58 \cdot \gamma_\lambda^{\;2} + 0,3}{4 \cdot \gamma_\lambda^{\;2} + 3,84} \quad , \; \lambda = 1,2,3,4 \quad ,$$

resulting in

$$s_{o_1} = s_{o_2} = + 0,042$$

$$s_{o_3} = s_{o_4} = - 0,35$$

Rule 17 : Departure of root locus branches from $s = 0$ (segment III, $k = k_{1_0}^j$)

The polynomial $R_{1o}(k) = R_o(k)$ has only complex conjugate roots. Therefore no branch meets the point $s = 0$.

Rule 2o : Return points

Relation (4.43) yields for $q = 2$

$$R(P,P_k') = Q_2 \cdot (Q_1 \cdot Q_1 - 4Q_o \cdot Q_2)$$

with roots

$$s_1 = - 0,05 \quad \text{and} \quad s_2 = - 0,16$$

as real return points for $k = 8,1$.

The root locus plots are drawn in Figures 5.4.b and 5.4.c .

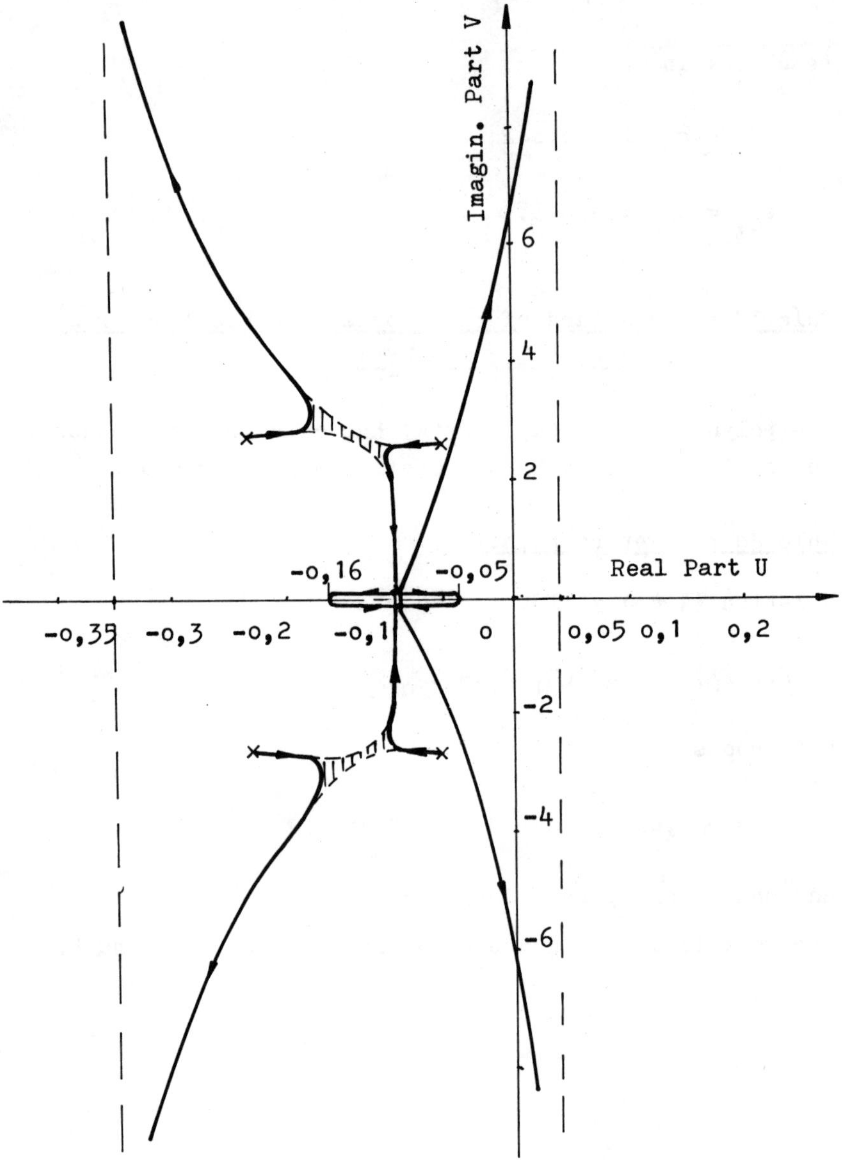

Figure 5.4.b: Qualitative root locus plot (a_o^2 = o.817).

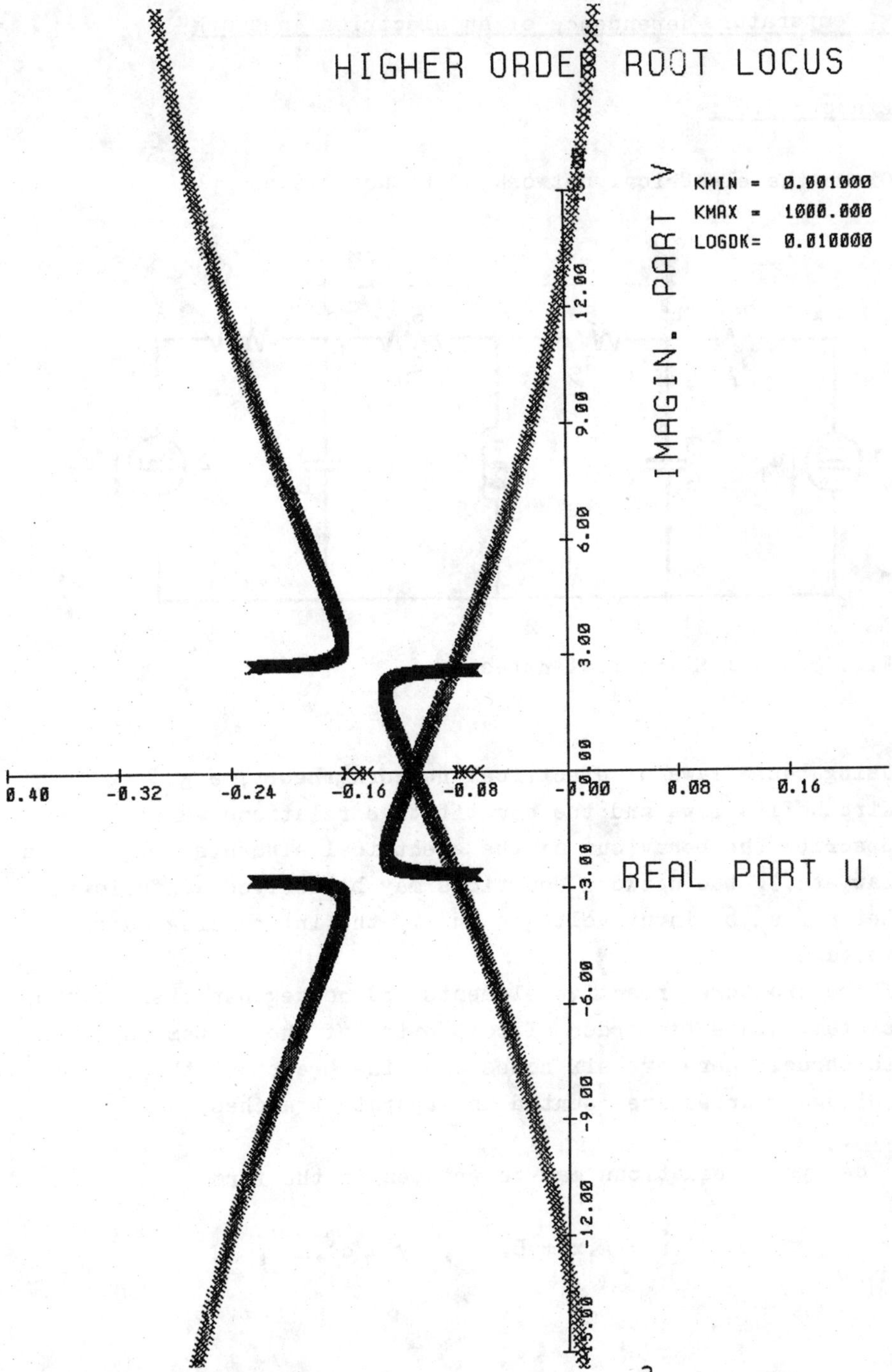

Figure 5.4.c: Computer root locus plot ($a_o^2 = 0.817$).

3. Temperature dependence of an electrical network

Example 5.5 :

Given the electrical network of Figure 5.5.a.

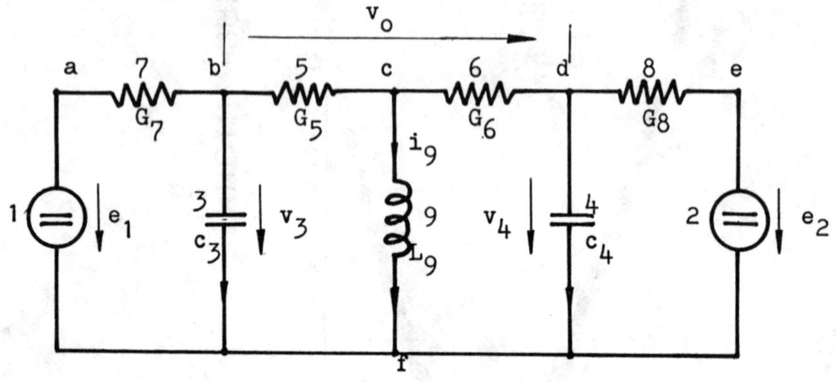

Fig. 5.5.a : Electrical network

Using basic laws of electrical network theory, e.g. Kirchhoff's laws and the constitutive relations which describe the behaviour of the electrical elements (e.g. Ohm's law etc.), the network equations may be derived as follows. Let e_1, e_2 be input voltages and v_o the interesting output voltage.
There are three reactive elements and no degeneracies in the system. Hence the order of complexity of the system is equal to three. There are six nodes and nine branches (the two voltage sources are counted as separate branches).

The system equations may be written in the form

$$\dot{x} = A.x + B.u \quad , \quad y = c^T.x \quad ,$$

where

$$x = \begin{pmatrix} v_3 \\ v_4 \\ i_9 \end{pmatrix} , \quad \bar{u} = \begin{pmatrix} e_1 \\ e_2 \end{pmatrix} , \quad y = v_0 ,$$

$$A = \begin{pmatrix} -\frac{G_7}{C_3} - \frac{G_5}{C_3} \cdot \frac{1}{N} , & \frac{G_5}{C_3} \cdot \frac{1}{N} , & \frac{G_5 \cdot R_6}{C_3} \cdot \frac{1}{N} \\ \frac{G_5}{C_4} \cdot \frac{1}{N} , & -\frac{G_8}{C_4} - \frac{G_5}{C_4} \cdot \frac{1}{N} , & \frac{1}{C_4} \cdot \frac{1}{N} \\ -\frac{G_5 \cdot R_6}{L_9} \cdot \frac{1}{N} , & -\frac{1}{L_9} \cdot \frac{1}{N} , & -\frac{R_6}{L_9} \cdot \frac{1}{N} \end{pmatrix} ,$$

$$B = \begin{pmatrix} -\frac{G_7}{C_3} , & 0 \\ 0 , & \frac{G_8}{C_4} \\ 0 , & 0 \end{pmatrix} , \quad C^T = (1, -1, 0); \; N := 1 + G_5 \cdot R_6 ,$$

or

$$A = \frac{1}{N} \cdot \begin{pmatrix} +\frac{1}{C_3} \cdot (-G_7 - G_7 \cdot G_5 \cdot R_6 - G_5) , & \frac{1}{C_3} \cdot G_5 & , \frac{1}{C_3} \cdot G_5 \cdot R_6 \\ \frac{1}{C_4} \cdot G_5 & , \frac{1}{C_4} \cdot (-G_8 - G_8 \cdot G_5 \cdot R_6 - G_5) , & \frac{1}{C_4} \\ -\frac{1}{L_9} \cdot G_5 \cdot R_6 & , -\frac{1}{L_9} & , -\frac{1}{L_9} \cdot R_6 \end{pmatrix} .$$

Let the resistive elements

$R_5 = R_6 = 1 \text{ k}\Omega$

$R_7 = R_8 = 2 \text{ k}\Omega$.

The temperature dependence of the reactive elements C_3, C_4 and L_9 is described as follows :

$$C_3 = C_4 = C_{30} \cdot \frac{T_o}{T_o + \alpha . T} ,$$

$$L_9 = L_{90} \cdot \frac{T_o}{T_o + \alpha . T} ,$$

$$k := \alpha . T \quad , \quad T_o = -273^oC \quad , \quad T = \text{temperature}$$

and

$$C_{30} = C_{40} = 1\,\mu F ,$$

$$L_{90} = 100 \text{ mH} .$$

Omitting the dimensions of the various coefficients, the system matrices take the form :

$$A = \frac{1}{2} \cdot \begin{pmatrix} -2.1o^{-3}/C_3 , & +1o^{-3}/C_3 , & 1/C_3 \\ 10^{-3}/C_4 , & -2.1o^{-3}/C_4 , & 1/C_4 \\ -\;1/L_9 , & -\;1/L_9 , & -1o^3/L_9 \end{pmatrix} ,$$

$$B = -\frac{5}{1o^4} \cdot \begin{pmatrix} -\;1/C_3 , & 0 \\ 0 , & -1/C_4 \\ 0 , & 0 \end{pmatrix} ,$$

$$c^T = (1, -1, 0) , \quad \text{and} \quad N = 2 .$$

The characteristic polynomial of matrix A takes the form :

$$\psi_A(s;k) = \frac{9.10^{-3}}{8.C_3^2.L_9} + \left(\frac{3}{2.C_3.L_9} + \frac{3.10^{-6}}{4.C_3^2}\right).s + \left(\frac{10^3}{2.L_9} + \frac{2.10^{-3}}{C_3}\right).s^2 + s^3$$

or

$$\psi_A(s;k) = 8.C_3^2.L_9.P(s;k) \quad , \quad \text{where}$$

$$P(s;k) = 9.10^{-3} + (12.C_3 + 6.L_9.10^{-6}).s$$

$$+ (4.C_3^2.10^3 + 16.10^{-3}.L_9.C_3).s^2 + 8.C_3^2.L_9.s^3 \quad ,$$

and using the expressions

$$C_3.C_4 = C_3^2 = C_{30}^2.T_o^2 \cdot \frac{1}{T_o^2 + 2.T_o.k + k^2} \quad ,$$

$$C_3.C_4.L_9 = C_{30}^2.L_{90}.T_o^3 \cdot \frac{1}{T_o^2 + 3.T_o^2.k + 3.T_o.k^2 + k^3} \quad ,$$

$$C_3.L_9 = C_{30}.L_{90}.T_o^2 \cdot \frac{1}{T_o^2 + 2.T_o.k + k^2}$$

results in

$$P(s;k) = \left\{ \frac{9.10^{-3}}{8.C_{30}^2.L_{90}.T_o^3} . \left[T_o^3 + 3.T_o^2.k + 3.T_o.k^2 + k^3 \right] \right\} .s^0$$

$$+ \left\{ \left[\frac{3}{2.C_{30}.L_{90}.T_o^2} + \frac{3.10^{-6}}{4.C_{30}^2.T_o^2} \right] . \left[T_o^2 + 2.T_o.k + k^2 \right] \right\} .s^1$$

$$+ \left\{ \left[\frac{10^3}{2.L_{90}.T_o} + \frac{2.10^{-3}}{C_{30}.T_o} \right] . \left[T_o + k \right] \right\} .s^2 \quad + \quad \{1\} .s^3$$

or

$P(s;k) =$

$$= \{(1,125.10^{10}).s^0+(15,75.10^6).s^1+(7.10^3).s^2+s^3\}.k^0$$

$$+ \{(1,236.10^8).s^0+(11,54.10^4).s^1+(2,56.10^1).s^2\}.k^1$$

$$+ \{(4,52.10^5).s^0+(2,113.10^2).s^1\}.k^2$$

$$+ \{(5,53.10^2).s^0\}.k^3$$

Construction of root locus plot.

Rule 1 : Polynomials $Q_j(s)$ and $R_i(k)$

	s^0	s^1	s^2	s^3	
Q_3	$0,553.10^3$				k^3
Q_2	$0,45.10^6$	$0,21.10^3$			k^2
Q_1	$0,124.10^9$	$0,115.10^6$	$0,256.10^2$		k^1
Q_0	$0,113.10^{11}$	$0,158.10^8$	$0,7.10^4$	1	k^0
	R_0	R_1	R_2	R_3	

Table 5.a.5 : Polynomial coefficients of $Q_j(s)$ and $R_i(k)$.

Rule 2 : Roots of $Q_j(s)$ and $R_i(k)$

Q_i	i	$s-s_1^i$	$s-s_2^i$	$s-s_3^i$	
Q_2	2	s+2110			k^2
Q_1	1	s+1780	s+2712		k^1
Q_0	0	s+2295	s+3148	s+1557	k^0
		R_0	R_1	R_2	

Table 5.b.5 : Roots of $Q_j(s)$

	s^0	s^1	s^2	
Q_2	k+212			$k-k_j^2$
Q_1	k+3oo +i.73.4	k+259		$k-k_j^1$
Q_0	k+3oo -i.73.4	k+288	k+272	$k-k_j^0$
j	o	1	2	
R_j	R_0	R_1	R_2	

Table 5.c.5 : Roots of $P_i(k)$

Rules 3 and 4 : Exponent Diagram

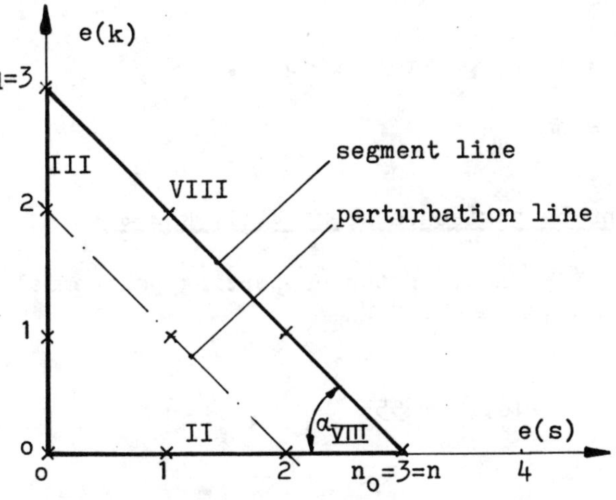

Figure 5.5.b : Exponent Diagram

Rule 5 : Constants

$$l_0 = 0 \ , \ n_0 = n_7 = 3 \ , \ q = 3$$

$$e_I = e_{IV} = e_V = e_{VII} = \bar{e}_{VI} = 0$$

$$e_{II} = e_{VIII} = 3 \ , \ \bar{e}_{III} = 3$$

Rule 7 : Slopes

$$\beta_{VIII} = 3/3 = 1$$

Note :

There exist three root locus branches starting at s_1^o, s_2^o and s_3^o , all of which tend towards infinity for increasing k. There exist three asymptotes and three finite asymptote points s_{oj} , $j = 1,2,3$.

Rule 11 : Angles of departure from s_j^o (segment II, k=o)

Replacing the variable s of $Q_o(s)$ and $Q_1(s)$ by (s'-2295), (s'-1557) and (s'-3148) and using (4.25b) results in

$$\begin{aligned} \varphi(-1557,o) &= \pi \\ \varphi(-2295,o) &= \pi \quad , \quad \text{for} \quad k \geq o \quad . \\ \varphi(-3148,o) &= \pi \end{aligned}$$

Rule 14 : Asymptote angles (segment VIII, k = ∞)

Using $\beta_{VIII} = 3/3 = 1$, the supporting polynomial of segment VIII

$$\gamma^3 + 25{,}6.\,\gamma^2 + 210.\gamma + 553 = o$$

with roots

$$\begin{aligned} \gamma_1 &= -\ 8{,}08 \\ \gamma_2 &= -\ 11{,}64 \quad , \\ \gamma_3 &= -\ 5{,}89 \end{aligned}$$

and relation (4.27b) yield

$$\varphi_1(\infty,\infty) = \pi$$

$$\varphi_2(\infty,\infty) = \pi \qquad \text{for} \quad k \geq 0 \; .$$

$$\varphi_3(\infty,\infty) = \pi$$

<u>Rule 15 :</u> <u>Asymptote points s_{oj}</u>

Beause of the results of Rule 14, the asymptote points have not to be computed.

<u>Rule 17 :</u> <u>Departure of root locus branches from s = o (segment III, $k = k_{1_o}^j$)</u>

No root locus branch meets the point, s = o for $k \geq 0$ (compare Table 5.c.5).

The root locus plots are drawn in Figures 5.5.c and 5.5.d .

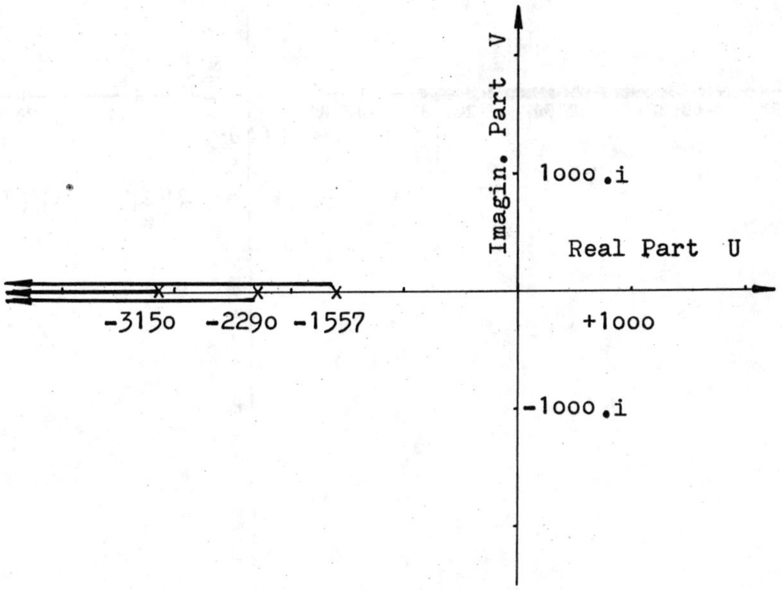

Figure 5.5.c: Qualitative root locus plot.

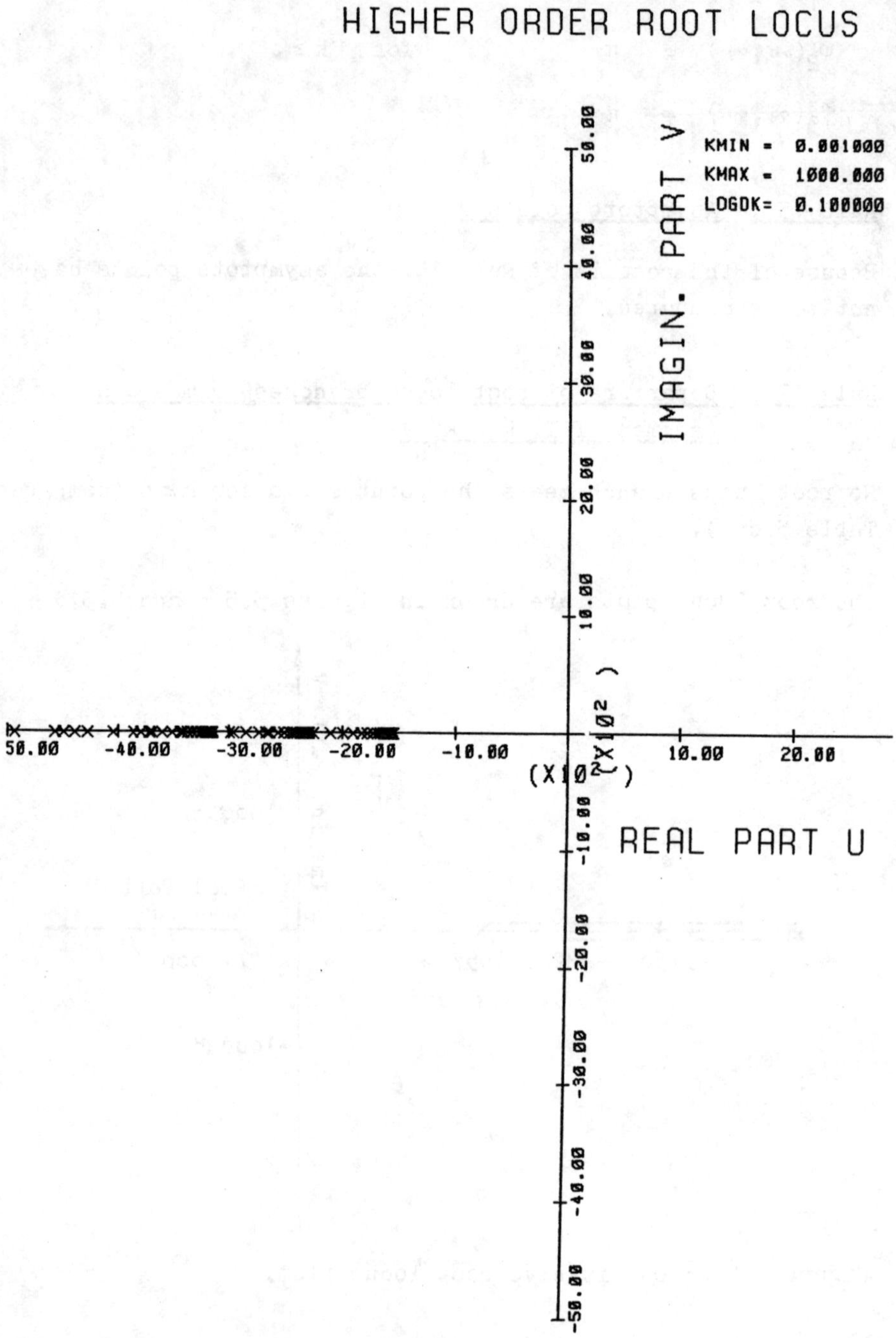

Figure 5.5.d: Computer root locus plot.

4. Linear multi-loop control system.

Example 5.6 : (Linear multi-loop control system).

Given a linear multi-loop control system of the form :

Open-loop system :

$$\dot{x} = A.x + B.u \ , \ y = C.x \ ,$$

Closed-loop system :

$$\dot{x} = (A - k.B.K_1).x + B.r \ , \ y = C.x \ ,$$

where

$$A = \begin{pmatrix} -1 & 1 & 2 \\ 0 & -2 & 1 \\ 0 & 0 & -3 \end{pmatrix} \quad , \quad B = C = I_3$$

and

$$K = k.K_1 = k.I_3 \qquad (I_3 = \text{ unit matrix}).$$

All state variables are fed back using a feedback factor k.

As is easily seen, the system is completely controllable and observable.

The transfer function matrix of the open loop system takes the form

$$P(s) = C.(sI - A)^{-1}.B$$

or

$$L(s) = P(s).K_1 = \begin{vmatrix} \frac{1}{(s+1)} & \frac{1}{(s+1).(s+2)} & \frac{2.s + 5}{(s+1).(s+2).(s+3)} \\ 0 & \frac{1}{(s+2)} & \frac{1}{(s+2).(s+3)} \\ 0 & 0 & \frac{1}{(s+3)} \end{vmatrix}$$

and

$$P(s) = \frac{1}{(s+1).(s+2).(s+3)} \begin{pmatrix} (s+2).(s+3) & & (2s+5) \\ 0 & (s+1).(s+3) & (s+1) \\ 0 & 0 & (s+1).(s+2) \end{pmatrix}$$

with the Mc Millan normal form

$$M_{Co}(s) = \begin{pmatrix} \frac{1}{(s+1).(s+2).(s+3)} & 0 & 0 \\ 0 & 1 & 0 \\ 0 & 0 & 1 \end{pmatrix} ,$$

the determinant divisors

$$D_0 = D_1 = 1 \quad , \quad D_2 = (s+1).(s+2).(s+3)$$

and the elementary divisors

$$f_0 = f_1 = 1, \; f_2 = f_3 = (s+1).(s+2).(s+3) \quad ,$$

where

$$D_i := \prod_{j=0}^{i} f_j \quad .$$

The transfer function matrix $P(s)$ is a <u>regular</u> matrix.

The <u>closed loop system</u> has the transfer function matrix

$$G_c(s;k) = \frac{1}{\Psi_c(s;k)} \cdot \begin{vmatrix} (z+2).(z+3) & (z+3) & (2z+5) \\ 0 & (z+1).(z+3) & (z+1) \\ 0 & 0 & (z+1).(z+2) \end{vmatrix}$$

and, using the abbreviation $z := s+k$,the related characteristic polynomial

$$\Psi_c(s;k) = (z + 1).(z + 2).(z + 3)$$

or

$$\psi_c(s;k) = (s^3+6.s^2+11.s+6)+k.(3.s^2+12.s+11)+k^2.(3.s+6)+k^3 \; .$$

The closed loop system has the Mc Millan normal form

$$M_{cc} = \begin{pmatrix} \frac{1}{\psi_c(s;k)} & 0 & 0 \\ 0 & 1 & 0 \\ 0 & 0 & 1 \end{pmatrix} .$$

Computation of the higher order root locus plot of the characteristic polynomial $\psi_c(s;k)$:

<u>Rule 1 : Polynomial coefficients of $Q_j(s)$ and $R_i(k)$; $j = 0, \ldots, 3$; $i = 0, \ldots 3$.</u>

	s^0	s^1	s^2	s^3	
Q_3	1				k^3
Q_2	6	3			k^2
Q_1	11	12	3		k^1
Q_0	6	11	6	1	k^0
	R_0	R_1	R_2	R_3	

Table 5.a.6 : Polynomial Coefficients of $Q_j(s)$ and $R_i(k)$.

Rule 2 : Roots of $Q_j(s)$ and $R_i(k)$

Q_j	i	$s-s_1^j$	$s-s_2^j$	$s-s_3^j$	
Q_2	2	(s+2)			k^2
Q_1	1	(s+1,4)	(s+2,6)		k^1
Q_0	o	(s+1)	(s+2)	(s+3)	k^0
		R_0	R_1	R_2	

Table 5.b.6: Roots of $Q_j(s)$.

	s^0	s^1	s^2	
Q_2	(k+3)			$k-k_j^2$
Q_1	(k+2)	(k+2,6)		$k-k_j^1$
Q_0	(k+1)	(k+1,4)	(k+2)	$k-k_j^0$
j	o	1	2	
R_j	R_0	R_1	R_2	

Table 5.c.6: Roots of $R_i(k)$

Rules 3 and 4 : Exponent Diagram

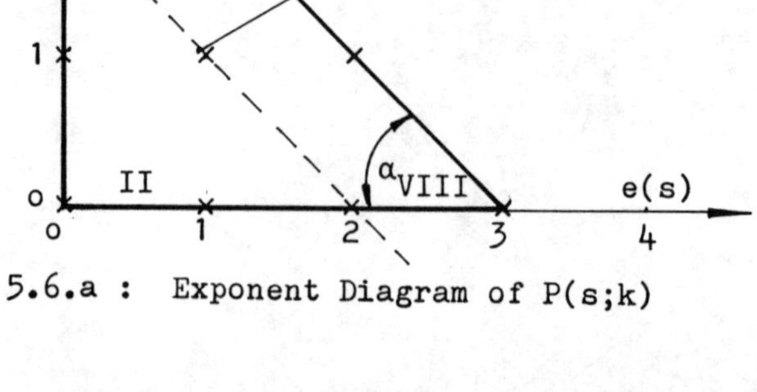

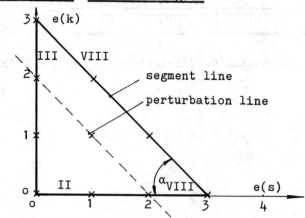

Figure 5.6.a : Exponent Diagram of P(s;k)

Rule 5 : Constants

$l_o = 0$, $n_o = n = 3$, $q = 3$

$e_{II} = 3$, $e_{VIII} = 3$, $\bar{e}_{III} = 3$

Rule 7 : Slopes

$\beta_{VIII} = 3/3 = 1$

Note :

There exist three root locus branches which tend to infinity for increasing values of k (k > o) and which start at the roots s_j^o of $Q_o(s)$; j = 1,2,3 .

Rule 11 : Angles of departure from $s_j^o \neq 0$ (segment II, k=o)

Replacing the variable s of $Q_o(s)$ and $Q_1(s)$ by the variables $s+s_j^o$ yields for

$s_1^o = -1$:

$$Q_o'(s') = 0 + 2.s' + \dots ,$$

$$Q_1'(s') = 2 + \dots ,$$

$$\gamma = -1 , \quad \beta_I = 1 ,$$

and

$$\varphi(-1,0) = \pi \qquad \text{for } k > 0 ,$$

for

$s_2^o = -2$:

$$Q_o'(s') = 0 - 1.s' + \dots$$

$$Q_1'(s') = -1 + \dots$$

$$\gamma = -1 , \quad \beta_I = 1$$

and

$$\varphi(-2,0) = \pi \qquad \text{for} \quad k \geq 0 \quad ,$$

and for

$s_3^0 = -3$:

$$Q_0'(s') = 0 + 2.s' + \ldots ,$$

$$Q_1'(s') = 2 + \ldots ,$$

$$\gamma = -1 \quad , \quad \beta_I = 1$$

and

$$\varphi(-3,0) = \pi \qquad \text{for} \quad k \geq 0 \quad .$$

Rule 14 : Asymptote angles (segment VIII, k =)

Using $\beta_{VIII} = 1$, the supporting polynomial

$$\gamma^3 + 3.\gamma^2 + 3.\gamma + 1 = 0$$

with roots

$$\gamma_1 = \gamma_2 = \gamma_3 = -1$$

and equation (4.27) results in

$$\varphi_1(\infty,\infty) = \varphi_2(\infty,\infty) = \varphi_3(\infty,\infty) = \pi$$

$$\text{for} \quad k > 0 \ .$$

Rule 15 : Asymptote points s_{oI}

Because of the results of Rule 14, the asymptote points have not to be computed.

Rule 17 : Departure of root locus branches from $s_j^o = 0$ for $k_o^j \neq 0$ (segment III, $k_o^j \neq 0$).

As is shown in Table 5.c.6, the point s = o is not met by

root locus branches for $k = k_o^j > 0$.

The root locus plot is shown in Figures 5.6.b and 5.6.c .

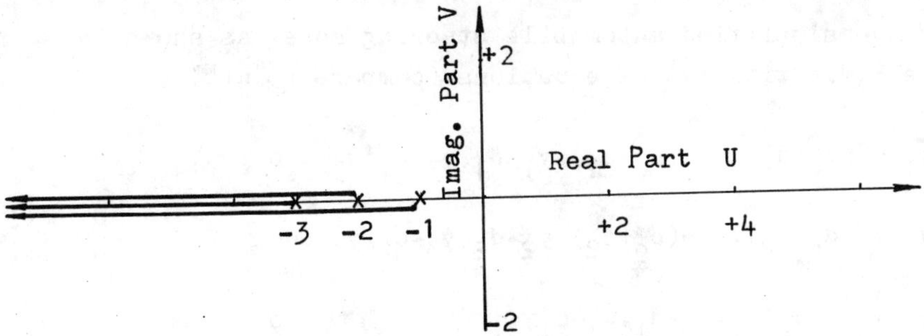

Figure 5.6.b: Qualitative root locus plot.

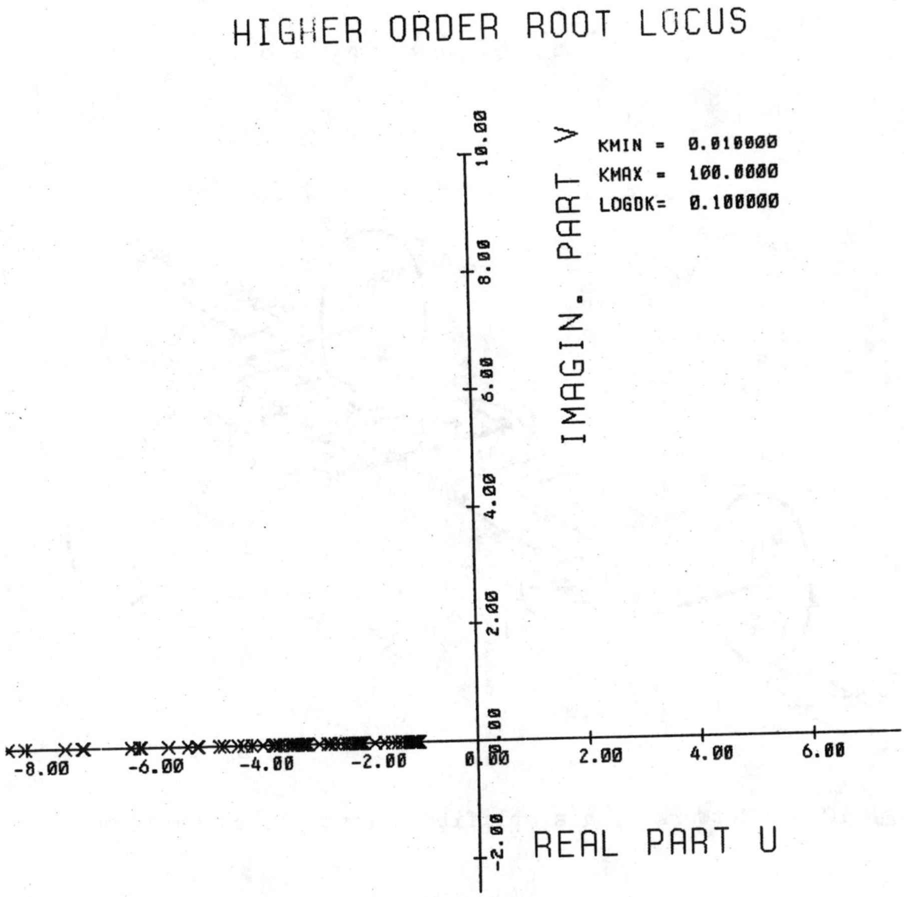

Figure 5.6.c: Computer root locus plot .

Example 5.7 : (Simple automobile steering model)

Given the simplified automobile steering model as shown in Figure 5.7.a with system equations (compare [5.4]) :

$$m_1 \cdot \ddot{y}_1 + (d_0+d_1) \cdot \dot{y}_1 + (c_0+c_1) \cdot y_1 - d_1 \cdot \dot{y}_3 - c_1 \cdot y_3 = 0$$

$$m_2 \cdot \ddot{y}_2 + (d_0+d_2) \cdot \dot{y}_2 + (c_0+c_2) \cdot y_2 - d_2 \cdot \dot{y}_3 - c_2 \cdot y_3 = 0$$

$$\Theta_2 \cdot \ddot{\varphi}_2 + d_1 \cdot \dot{\varphi}_2 + c_1 \cdot \varphi_2 - d_1 \cdot \dot{\varphi}_1 - c_1 \cdot \varphi_1 = 0$$

$$m_3' \cdot \ddot{y}_3 + d_3' \cdot \dot{y}_3 + c_3' \cdot y_3 - d_1 \cdot \dot{y}_1 - d_2 \cdot \dot{y}_2 - c_1 \cdot y_1 - c_2 \cdot y_2 - \ddot{u} \cdot d_1 \cdot \dot{\varphi}_2 - \ddot{u} \cdot c_1 \cdot \varphi_2 = 0 \quad ,$$

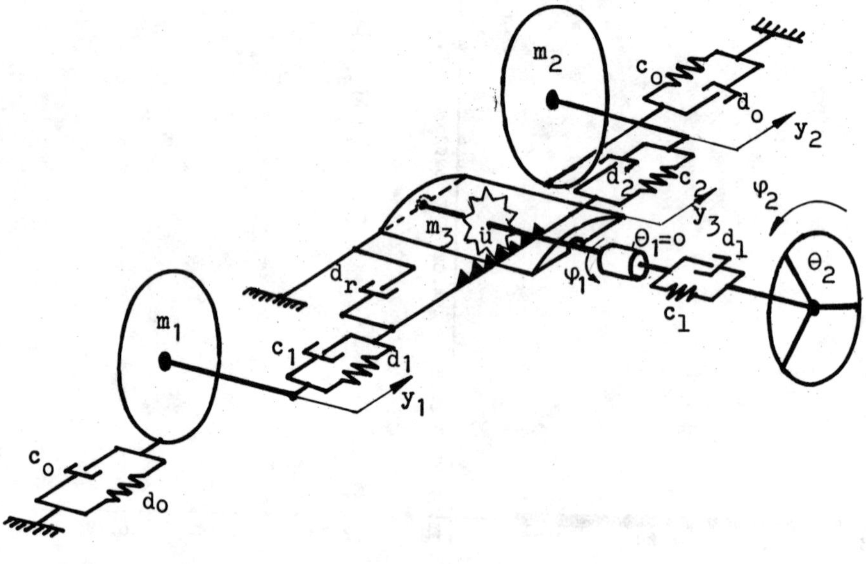

Fig. 5.7.a: Network of a simplified automobile steering model.

where

$$m_3' := (m_3 + \ddot{u}.\ \Theta_1)$$

$$d_3' := (d_r + d_1 + d_2 + \ddot{u}^2.d_1)$$

$$c_3' := (c_1 + c_2 + \ddot{u}^2.c_1)\ .$$

Using the relations

$$c_1 = c_2\ ,\ d_1 = d_2\ ,\ \varphi_1 = \ddot{u}.y_3\ ,$$

$$x_2 := y_1\ ,$$
$$x_4 := y_3\ ,$$
$$x_6 := y_2\ ,$$
$$x_8 := \varphi_2\ ,$$

and

$$a_{11} = a_{55} = -(d_o + d_1)/m_1 = 3{,}5$$

$$a_{12} = a_{56} = -c_o/m_1 - c_1/m_1 = -10^4 - 2500.k\ ,$$

$$a_{13} = d_1/m_1 = 1{,}0\ ,$$

$$a_{14} = a_{54} = c_1/m_1 = 2500k\ ,$$

$$a_{21} = 1\ ,$$

$$a_{33} = -d_3'/m_3' = 1960\ ,$$

$$a_{31} = d_1/m_3' = 20\ ,$$

$$a_{32} = a_{36} = c_1/m_3' = 50000.k\ ,$$

$$a_{34} = -\ddot{u}^2.c_1/m_3' - 2c_1/m_3' = -6{,}656.10^5 - 10^5.k\ ,$$

$$a_{35} = d_2/m_3' = 20\ ,$$

$$a_{37} = \ddot{u}.d_1/m_3' = 3{,}5.10^5\ ,$$

$$a_{38} = \ddot{u}.c_1/m_3' = 4150\ ,$$

$$a_{43} = 1 \quad ,$$

$$a_{53} = d_2/m_2 \quad ,$$

$$a_{65} = 1 \quad ,$$

$$a_{73} = ü . d_1/\Theta_2 = 480 \quad ,$$

$$a_{74} = ü . c_1/\Theta_2 = 1{,}60640 . 10^5 \quad ,$$

$$a_{77} = - d_1/\Theta_2 = 3 \quad ,$$

$$a_{78} = - c_1/\Theta_2 = 1{,}04 . 10^3 \quad ,$$

$$a_{87} = 1 \quad ,$$

$$c_1 = c_2 = 10^5 . k \quad ,$$

where the dimension of the coefficients is omitted, the following state equations may be formulated :

$$\dot{x} = A_0 . x + k . A_1 . x \quad ,$$

where $x = [x_1 , x_2 , x_3 , x_4 , x_5 , x_6 , x_7 , x_8]^T$,

$$A_0 = \begin{pmatrix} -3{,}5 & -15000 & +1 & +5000 & 0 & 0 & 0 & 0 \\ 1 & 0 & 0 & 0 & 0 & 0 & 0 & 0 \\ +20 & 100000 & -1960 & -865600 & +20 & 100000 & +12 & 4160 \\ 0 & 0 & 1 & 0 & 0 & 0 & 0 & 0 \\ 0 & 0 & 1 & +5000 & -3{,}5 & -15000 & 0 & 0 \\ 0 & 0 & 0 & 0 & 1 & 0 & 0 & 0 \\ 0 & 0 & 480 & 166400 & 0 & 0 & -3 & -1040 \\ 0 & 0 & 0 & 0 & 0 & 0 & 1 & 0 \end{pmatrix}$$

and

$$A_1 = \begin{pmatrix} 0 & -1000 & 0 & +1000 & 0 & 0 & 0 & 0 \\ 0 & 0 & 0 & 0 & 0 & 0 & 0 & 0 \\ 0 & 2000 & 0 & -40000 & 0 & 20000 & 0 & 0 \\ 0 & 0 & 0 & 0 & 0 & 0 & 0 & 0 \\ 0 & 0 & 0 & 1000 & 0 & -1000 & 0 & 0 \\ 0 & 0 & 0 & 0 & 0 & 0 & 0 & 0 \\ 0 & 0 & 0 & 0 & 0 & 0 & 0 & 0 \\ 0 & 0 & 0 & 0 & 0 & 0 & 0 & 0 \end{pmatrix} .$$

The characteristic polynomial of the above state space equation with $k = 10^{-5}.c_1 = 10^{-5}.c_2$ as parameter takes the form as shown in Table 5.a.7 .

Rule 1 : Polynomials $Q_i(s)$ and $R_j(k)$.

The coefficients of the polynomials $Q_i(s)$ and $R_j(k)$ are collected in Table 5.a.7 .

Rule 2 : Roots of $Q_i(s)$ and $R_j(k)$.

The roots of the polynomials $Q_i(s)$ are listed in Table 5.b.7.

The roots of the polynomials $R_j(k)$ are collected in Table 5.c.7 .

	s^0	s^1	s^2	s^3	s^4	s^5	s^6	
Q_1	k+15,2	k+15,6	k+15,9	k+20,1	k+51,9			$k-k_j^1$
Q_0	k+4,97	k+5,35	k+10,5	k+12,5	k+12,1	k+15,1	k+21,6	$k-k_j^0$
j	0	1	2	3	4	5	6	
R_j	R_0	R_1	R_2	R_3	R_4	R_5	R_6	

Table 5.c.7 : Roots of $R_j(k)$; $j = 0,1, \ldots , 6$.

	s^0	s^1	s^2	s^3	s^4	s^5	s^6	s^7	s^8	
Q_2	4,132o59 .1o^{14}	1,32977 .1o^{12}	1,1o743 .1o^{12}	2,148o3 .1o^{9}	4,o9743 .1o^{7}					k^2
Q_1	8,32672 .1o^{15}	2,79o68 .1o^{13}	2,92597 .1o^{13}	7,oo372 .1o^{10}	2,61954 .1o^{9}	4,25191 .1o^{6}	42ooo,			k^1
Q_0	3,11947 .1o^{16}	1,11343 .1o^{14}	1,85261 .1o^{14}	5,4o485 .1o^{11}	2,5654 .1o^{10}	6,53286 .1o^{7}	9,1o474 .1o^{5}	197o,	1,o	k^0
	R_0	R_1	R_2	R_3	R_4	R_5	R_6	R_7	R_8	

Table 5.a.7 : Coefficients of Polynomials $Q_j(s)$ and $R_i(k)$.

Q_i	i	$s-s_1^i$	$s-s_2^i$	$s-s_3^i$	$s-s_4^i$	$s-s_5^i$	$s-s_6^i$	$s-s_7^i$	$s-s_8^i$	
Q_2	2	s+25,97 +i.161,1	s+25,97 -i.161,1	s+o,24o +i.19,46	s+o,24o -i.19,46					k^2
Q_1	1	s+47,89 +i.211,8	s+47,89 -i.211,8	s+2,586 +i.119,8	s+2,586 -i.119,8	s+o,141 +i.17,9	s+o,141 -i.17,9			k^1
Q_0	o	s+667,3	s+1293	s+1,752 +i.122,5	s+1,752 -i.122,5	s+2,835 +i.118,1	s+2,835 -i.118,1	s+o,o568 +i.13,13	s+o,o568 -i.13,13	k^0
		R_0	R_1	R_2	R_3	R_4	R_5	R_6	R_7	

Table 5.b.7 : Roots of $Q_i(s)$; i = o,1,2 .

Rules 3 and 4 : Exponent Diagram

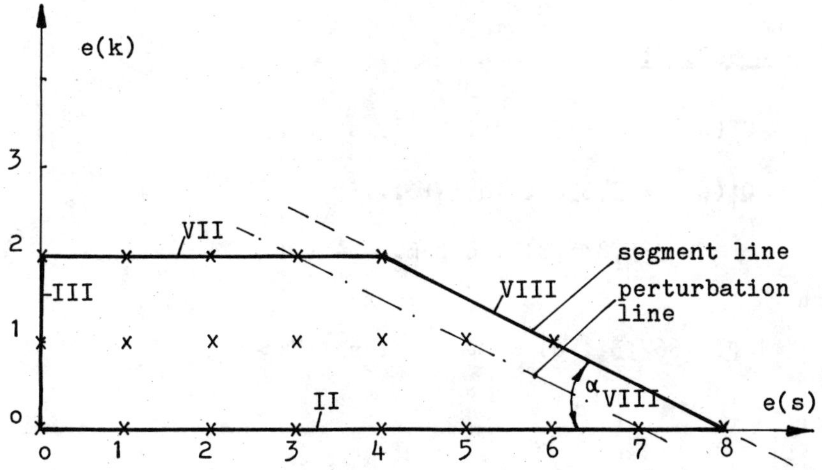

Fig. 5.7.b : Exponent Diagram .

Rule 5 : Constants

$$l_o = 0 \text{ , } n_o = n = 8 \text{ , } q = 2$$

$$\bar{e}_{III} = 2 \text{ , } e_{II} = 8 \text{ , } e_{VII} = 4 \text{ , } e_{VIII} = 4$$

Rule 7 : Slopes

$$\beta_{VIII} = 2/4 = 0{,}5 \quad .$$

Note :

In agreement with the above stated,rules there exist $n = 8$ root locus branches which start at the roots s_1^o of $Q_o(s)$ for $k = 0$. Four of these branches tend towards the sinks s_i^2 of $Q_2(s)$ for $k = \infty$. The other four branches tend to infinity. No branch meets the point $s = 0$ for positive real values of k.

Rule 1o : Angles of departure from $s_j^o = 0, j = 1, \ldots, 8$ (segment II, k = o).

Replacing the variable s of $Q_o(s)$ and $Q_1(s)$ by the variables $s' + s_j^o$ yields for

$s_1^o = -667,3$:

$$Q_o'(s') = 0 + 5,817 \cdot 1o^{19} \cdot s' + \ldots$$

$$Q_1'(s') = 3,658 \cdot 1o^{21} + \ldots$$

$$\gamma = -365,8/5,817 > 0 \;, \quad \beta_I = \frac{1}{1} = 1$$

and

$$\varphi(-667,3 \; ; \; o) = \pi \qquad \text{for} \quad k > o \;;$$

for

$s_2^o = -1293,4$:

$$Q_o'(s') = 0 - 2,96 \cdot 1o^{21} \cdot s' + \ldots$$

$$Q_1'(s') = +1,885 \cdot 1o^{23} + \ldots$$

$$\gamma = 188,5/2,96 > 0 \;, \quad \beta_I = 1 \quad \text{and}$$

$$\varphi(-1293,4 \; ; \; o) = o \qquad \text{for} \quad k > o \;;$$

for

$s_3^o = -2,835 + i.\,118,1$:

$$Q_o'(s') = 0 + (2,28 - i.30,5).1o^{14} \cdot s' + \ldots$$

$$Q_1'(s') = (-8,01 - i.1,76).1o^{15} + \ldots$$

$$\gamma = 10.(8,01 + i.1,76)/(2,28 - i.30,5) \;,$$

$$\arg \gamma \cong 98^o \;, \quad \beta_I = 1 \quad \text{and}$$

$$\varphi(-2,835 + i.118,1 \; ; \; o) = 98^o \qquad \text{for} \quad k > o \;,$$

and

$$(-2{,}835 - i.118{,}1\ ;\ 0) = -98^o \quad \text{for} \quad k > 0\ ,$$

for

$\underline{s_5^0 = -1{,}753 - i.\ 122{,}46:}$

$$Q_0'(s') = 0 - (1{,}89 + i.34{,}2).10^4.s' + \ldots\ ,$$

$$Q_1'(s') = (13{,}9 + i 0{,}75)\ .10^{15} + \ldots\ ,$$

$$\gamma = 10^{11}.(13{,}9 + i 0{,}75)/(1{,}89 + i.34{,}2)\ ,$$

$$\arg\gamma = -82^o\ , \quad \beta_I = 1 \quad \text{and}$$

$$\varphi(-1{,}753 - i.122{,}46\ ;\ 0) = -82^o \quad \text{for} \quad k > 0$$

and

$$\varphi(-1{,}753 + i.122{,}46\ ;\ 0) = +82^o \quad \text{for} \quad k > 0\ ,$$

for

$\underline{s_7^0 = -0{,}0568 + 13{,}14:}$

$$Q_0'(s') = 0 + (-1{,}77 + i.46{,}34).10^{14}.s' + \ldots\ ,$$

$$Q_1'(s') = (33{,}57 + i.1{,}67).10^{14} + \ldots\ ,$$

$$\gamma = (33{,}57 + i.1{,}67)/(1{,}77 - i.46{,}34)\ ,$$

$$\arg\gamma = 90^o\ , \quad \beta_I = 1 \quad \text{and}$$

$$\varphi(-0{,}0568 + i.13{,}14\ ;\ 0) = 90^o \quad \text{for} \quad k > 0$$

and

$$\varphi(-0{,}0568 - i.13{,}14\ ;\ 0) = -90^o \quad \text{for} \quad k > 0\ ,$$

for

Rule 13 : Angles of arrival at $s_j^2 \neq o$ (segment VII;k = ∞).

Replacing the variable s of $Q_2(s)$ and $Q_1(s)$ by the variables $s' - s_j^2$ yields for

$s_1^2 = -25{,}97 + i.161{,}1$:

$$Q_2'(s') = o - (2{,}684 + i.1{,}086).10^9 .\ s' + \dots ,$$

$$Q_1'(s') = (0{,}9075 + i.2{,}917).\ 10^{17} + \dots ,$$

$$\gamma = 10^8.(0{,}9075 + i.2{,}917)/(2{,}684 + i.1{,}086) ,$$

$$\arg\gamma = 40^\circ , \quad \beta_I = 1 \quad \text{and}$$

$$\varphi(-25{,}97 + i.161{,}1\ ; \infty) = 130^\circ \quad \text{for} \quad k > o$$

and

$$\varphi(-25{,}97 - i.161{,}1\ ; \infty) = -130^\circ \quad \text{for} \quad k > o ,$$

for

$s_3^2 = -0{,}24 + 19{,}4$:

$$Q_2'(s') = o + (-0{,}1597 + i.4{,}18).10^{13}.s' + \dots ,$$

$$Q_1'(s') = 8{,}326.10^{15} + \dots ,$$

$$\gamma = 832{,}6/(-0{,}1597 + i.4{,}18)$$

$$\arg\gamma = 67^\circ , \quad \beta_I = 1 \quad \text{and}$$

$$\varphi(-0{,}24 + i.19{,}4\ ; \infty) = 157^\circ \quad \text{for} \quad k > o$$

and

$$\varphi(-0{,}24 - i.19{,}4\ ; \infty) = -157^\circ \quad \text{for} \quad k > o .$$

Rule 14 : Asymptote angles (segment VIII , k = ∞).

Using $\beta_{VIII} = 1/2$, the supporting polynomial of segment VIII

$$\gamma^4 + 4{,}2.10^4.\gamma^2 + 4{,}1.10^7 = 0$$

with roots

$$\gamma_1 = - i.202{,}49$$

$$\gamma_2 = + i.202{,}49$$

$$\gamma_3 = - i.31{,}61$$

$$\gamma_4 = + i.31{,}61$$

yields the asymptote angles (compare 4.27b) :

$$\begin{aligned} \varphi_1(\infty,\infty) &= -\pi \\ \varphi_2(\infty,\infty) &= +\pi \\ \varphi_3(\infty,\infty) &= -\pi \\ \varphi_4(\infty,\infty) &= +\pi \end{aligned} \qquad (\text{for } k > 0) \quad .$$

<u>Rule 15 : Asymptote points s_{oj} .</u>

Using relation (4.37a) in connection with Figure 5.7.b and the roots $\gamma_j(j = 1,2,3,4)$ of the supporting polynomial related to segment VIII yields the asymptote points

$$s_{o_j} = \frac{1970.\gamma_j^7+4{,}25.10^6.\gamma_j^5+2{,}15.10^9.\gamma_j^3}{8.\gamma_j^7+2{,}6.10^5.\gamma_j^5+1{,}639.10^8\gamma_j^3} \quad .$$

$$j = 1,2,3,4 \; ,$$

or

$$s_{o_1} = s_{o_3} = - \; 1{,}5$$

$$s_{o_2} = s_{o_4} = - \; 1100 \quad .$$

Rule 17 : Departure of root locus branches from $s_j^o = o$ for $o < k < \infty$ (segment III).

As is shown in Table 5.c.7, the polynomial $R_o(k)$ has no positive real root. Therefore no branch meets the point s = o for positive real values of k.

The root locus plots are drawn in Figures 5.7.c , 5.7.d and 5.7.e .

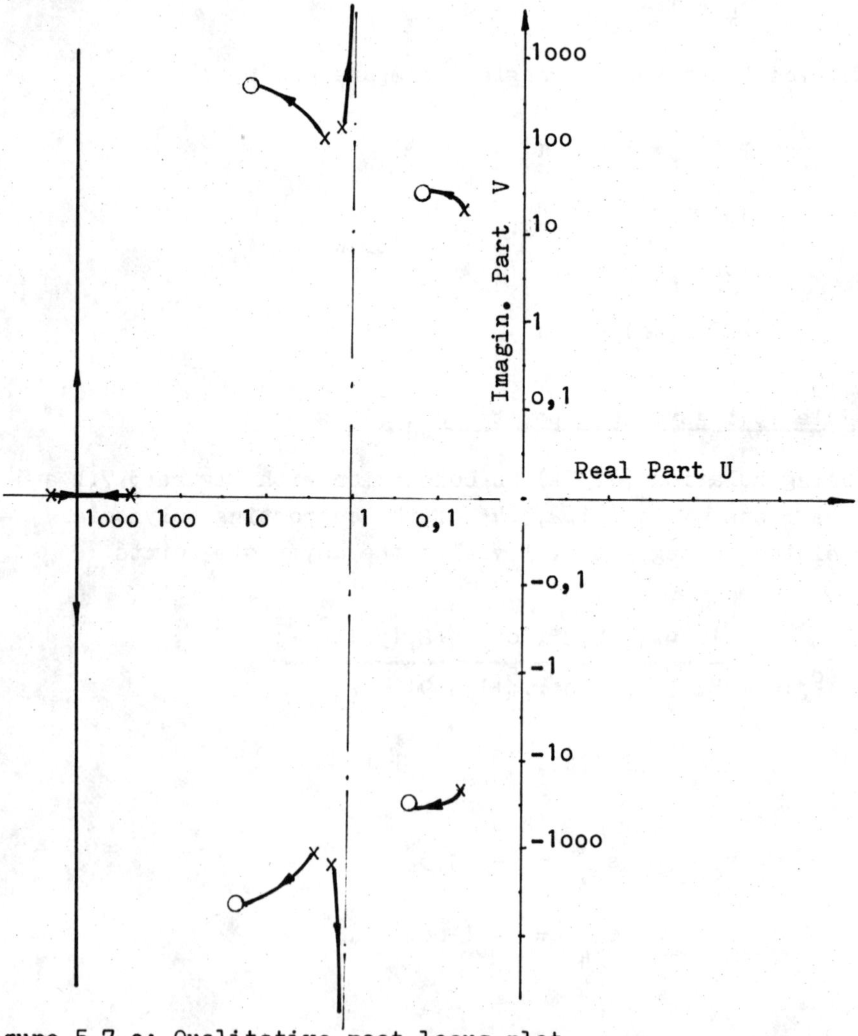

Figure 5.7.c: Qualitative root locus plot .

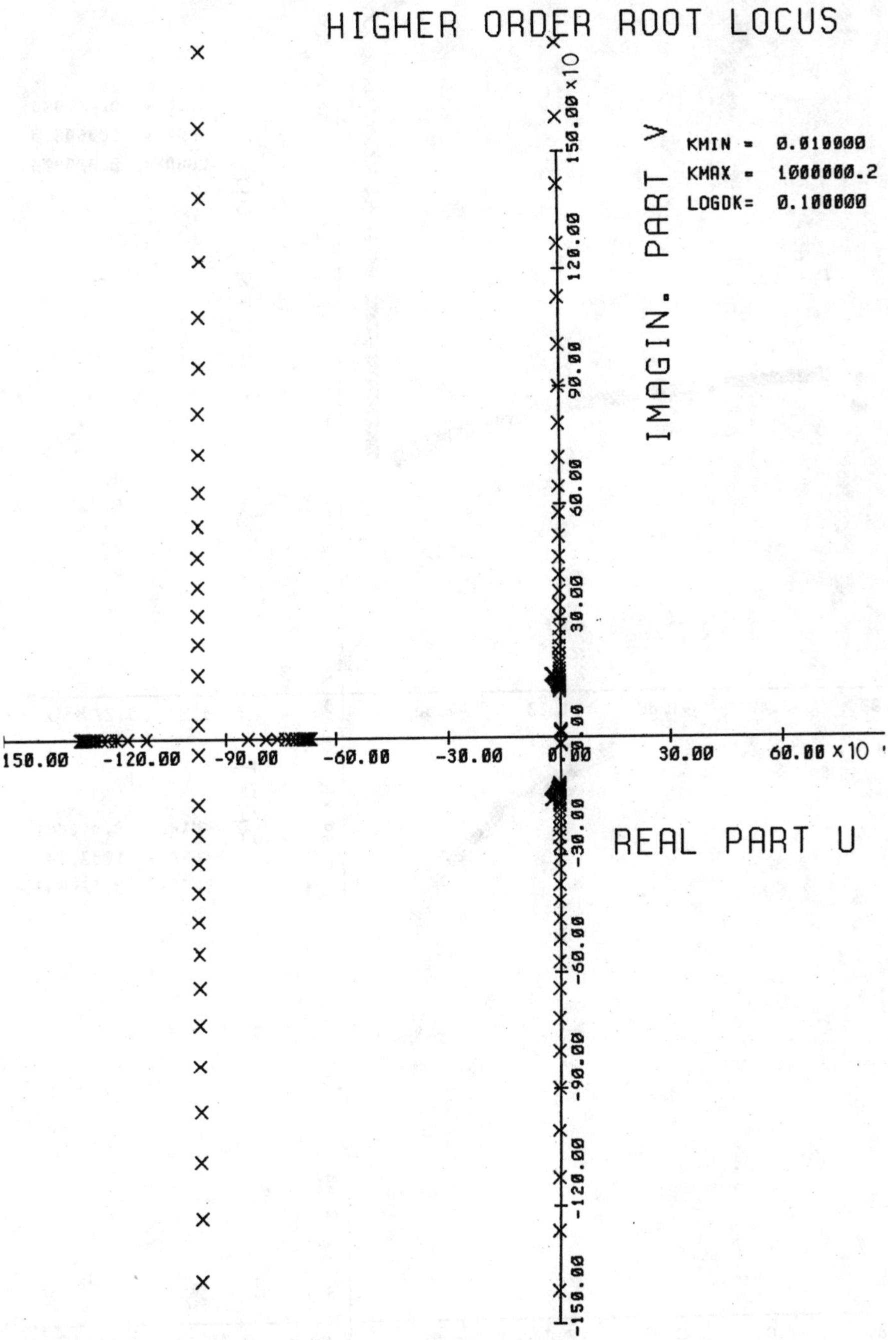

Figure 5.7.d: Computer root locus plot.

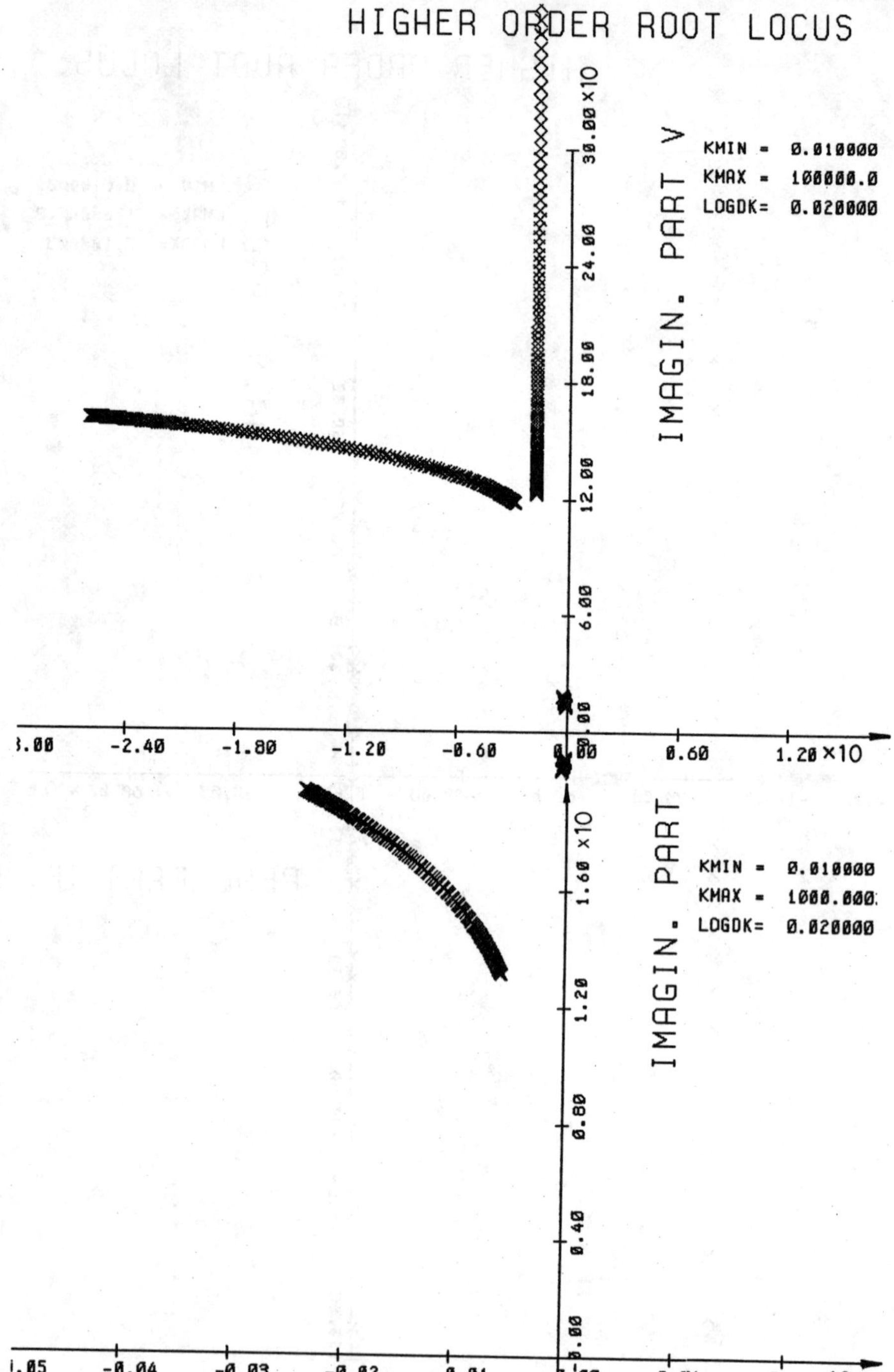

Figure 5.7.e: Computer root locus plot.

References

[5.1] H. Hahn, Zur Theorie und Technik singulärer Regelkreise, Habilitationsschrift, Universität Tübingen, Fachbereich Physik, 1977/78, Chapter VI .

[5.2] J.H. Blakelock, Automatic Control of Aircraft and Missiles, J. Wiley, 1965.

[5.3] G. Rosenau, Höhere Wurzelortskurven bei Mehrfachregelsystemen, IFAC Symposium, Düsseldorf, 1968.

[5.4] H. Hahn et. al. ,Simulationsmodelle für Zahnradlenkungen, BMW-Bericht E21-3o/78, 1978.

VI. Comparison of Classical Versus Higher Order Root Locus Techniques and Simple Design Steps.

In this Chapter VI.1, the classical and the higher order root locus techniques are compared to each other from the geometrical as well as from the structural point of view. A set of geometrical peculiarities which can only appear in higher order root locus plots is collected. Some of these geometric phenomena as well as the close and direct correlation between specific points and segments of the exponent diagram and special features of the root locus pattern are used for synthesis purposes in Chapter VI.2 . Various examples presented in Chapter V are modified on the basis of these relations.

1. Comparison of classical versus higher order root locus techniques

Both, classical as well as higher order root locus techniques are trial and error synthesis tools. Their common aim is to provide insight into the system behaviour and to deliver strong hints how to systematically improve a concrete system design. These techniques provide both, simple rules for getting a qualitative and rough sketch of the root locus plot and quantitative results by using a digital computer. Therefore they are to be settled among the various methods called computer aided design procedures.

The differences between the classical and the higher order root locus techniques may be discussed from various points of view.

A comparison of the polynomials (4.1) and (4.2) of Chapter IV shows the formal differences between the characteristic polynomials which may be investigated by the two methods. Classical root locus technique treates polynomials which are linear in the parameter k (compare (4.1)), whereas polynomials which depend nonlinear on a scalar parameter k (compare (4.2) to (4.6)) can be discussed by the higher order root locus technique. Clearly, equation (4.1) is a special case of equations (4.2) to (4.6).

Classical root locus technique is useful for the design of single-loop control systems and for the sensitivity analysis of simple active and passive systems (control systems, electrical, mechanical, hydraulic, pneumatic and thermodynamic networks).
More complex situations like multiloop control systems and large scale mechanical and electrical systems usually have characteristic polynomials of the form of equations (4.2) to (4.6). They must be treated by higher order root locus techniques. Some examples of this kind from the field of practical applications are collected in Chapter V.

From the geometrical point of view, the differences between the classical and the higher order root locus techniques may be characterized by looking through different types of glasses at a dynamic system.
The classical root locus technique may be symbolized by glasses with windows, each of which has at most four edges (compare Fig. 6.o.a), whereas the higher order root locus technique is represented by windows with n edges, n arbitrary large but finite (compare Fig. 6.o.b).

In the framework of this allegory, the glasses stand for the two types of exponent diagrams related to classical and to higher order root locus techniques, respectively.
Exponent diagrams corresponding to classical root locus technique are shown in Figure 4.1 of Chapter IV. They embrace at most four of the segments I to VIII simultaneously, and each of the segments occuring consists of at most one single straight line.
A general exponent diagram related to the higher order root locus technique is shown in Figures 3.1 and 3.2 of Chapter III. It may comprehend all of the eight types of segments at the same time, where in addition some of these segments may be polygons consisting of various straight lines.

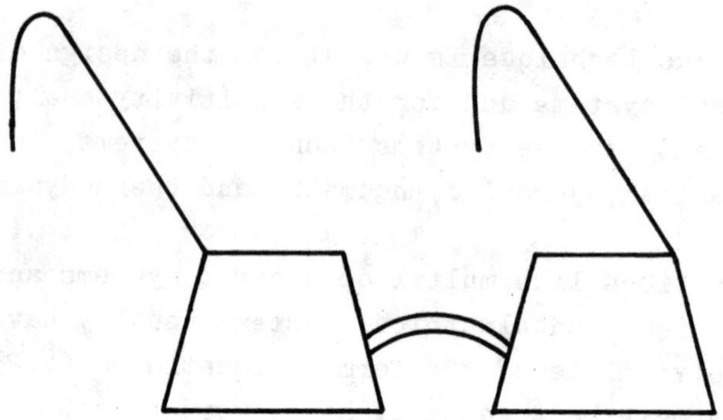

Fig. 6.o.a: "First order root locus glasses".

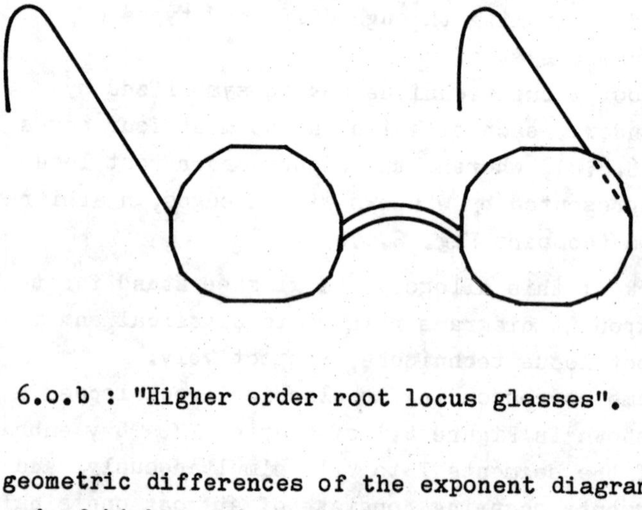

Fig. 6.o.b : "Higher order root locus glasses".

The geometric differences of the exponent diagrams of classical and of higher order root locus techniques indicate already obviously the different <u>order of complexity</u> of the two methods. Even more, the geometric structure of this pattern announces basic elements of the internal structure of the root locus plot. It allows a concrete control theoretical interpretation of various elements of the exponent

diagram, and it may give concrete hints for improving a system design (compare Chapter VI.2).
Each of the segments I to VIII of the exponent diagram embraces a set of straight lines of similar position within the exponent polygon. The eight segments differ qualitatively from each other.

The number and type of segments occuring in the exponent diagram of a system give rise to what will be called the "coarse structure" of the related root locus plot. As has been pointed out in the previous chapters (Chapters III, IV and V), each of the eight segments has a specific control - or system theoretical interpretation. As the exponent diagram of a single loop control system (with k as feedback factor) contains at most four of the eight segments simultaneously, whereas the exponent diagram of a multiloop control system may comprehend all eight types of segments at the same time, the coarse structure of the root locus plot of a multi-loop system is much richer then the one of a single loop control system.

Four segments (types I, IV, V and VIII) of the exponent diagram may consist of a polygon of arbitrary (but finite) length.
The number of straight lines of each segment constitutes, what will be called a "fine-structure" of the related root locus plot. As has been shown in Chapter IV, Figure 4.1, each segment of a single loop control system (with k as feedback factor) consists of exactly one straight line. Therefore, the related root locus plots don't have a fine-structure, contrary to multiloop systems.

Each straight line of a segment of the exponent polygon may meet two or more points of the exponent diagram. Segments of the latter kind constitute what will be called a "hyper-fine-structure" of the related root locus plot in what follows. Again, classical root locus plots don't provide a hyper-fine-structure. As has been shown in Chapters III, IV and V, the coarse structure of the exponent diagram defines

the basic geometrical elements which constitute the related root locus pattern as for instance the occurence and number of asymptotes and system zeros or sinks and so on (compare Chapter IV). The <u>fine-structure</u> of the exponent diagram creates subgroups or subsets of a set of geometric root locus elements of a given kind. It divides, for instance, the set of asymptotes into various groups, each of which constitutes a specific geometric pattern of the root locus plot. The same holds for the set of angles of departure of root locus branches from roots of $Q_o(s)$ or for the set of angles of arrival of root locus branches at sinks, and so on.
The <u>hyper-fine-structure</u> of the exponent diagram determines the geometric features of the above defined subgroups or subsets of geometric elements. If, for instance, a straight line of segment VIII meets exactly two points of the exponent diagram, the related subset of asymptotes constitutes a <u>Butterworth Configuration</u>. If this straight line meets more than two points of the exponent diagram, the related subset of asymptotes is no Butterworth Configuration. This feature may be of basic importance in control system design, as will be pointed out in Chapter VI.2 .

Therefore, the pattern of all asymptotes may consist of various stars of the Butterworth type,and of other stars which are not of this type. Clearly, the set of asymptotes of classical root locus plots always defines one single Butterworth Configuration.
Similar results hold for the other segments of the exponent diagram (compare Chapters IV and V).

The increased complexity of the structure of higher order root locus pattern compared to classical root locus plots implies that there is a need for more and more complex root locus construction rules than in the classical case (compare Chapter IV). Though, in specific situations, for example in case of asymptotes which build a Butterworth Configuration, the construction rules for asymptote angles as well as for asymptote points reduce to the related well known rules of Evans. This has been shown in Chapter IV, equations (4.29) to

(4.35) and (4.4o).

The geometric counterpart of the increased algebraic complexity of the characteristic polynomial is the occurence of new geometric phenomena of the higher order root locus plots compared to the classical root locus plots.
Without going into detail, only a few of these phenomena will be mentiowed here,some of which will turn out to be of some importance in system design.

The nonlinear dependence of equation (4.2) on the parameter k allows the occurence of return-points of root locus branches on the real axis.

As a consequence, parts of the real axis of the complex plane may be covered several times by root locus branches (multiple covering of parts of the real axis).

Specific points of the complex plane may be met twice or more by one root locus branch or by various root locus branches. A special case of this phenomenon is called "loop fixation" in Chapter VII.

The occurence of pattern of asymptotes which don't constitute a Butterworth Configuration has already been mentioned.

The roots of the polynomial $Q_q(s)$ or $Q_{r_q}(s)$ (compare Figures 3.1 and 3.2 of Chapter III and equation (4.5) of Chapter IV) are called sinks of the root locus plot and of the related system (compare Chapters IV, V and VII). The sinks are the final goals of root locus branches, the final points of the complex plane,where root locus branches end for $k = \infty$. In some situations, the sinks may be interpreted as generalized zeros of linear or nonlinear multi-loop control systems. Contrary to linear single-loop control systems, where the root locus sinks are identical to the zeros of the system transfer function, the sinks of linear multi-loop control systems may differ from the zeros of the transfer function matrix in the sense of McMillan (compare Capter VII).
The above mentioned geometric phenomena of higher order root locus plots may appear,at the first glance,as abstract

curiosities or as exotic phenomena which are useless in practical applications. But the contrary is the case, as will be shown in Chapter VI.2 .

2. Simple design steps.

According to Chapter IV, there exists a strong correlation between specific points and segments of the exponent diagram on one hand , and between characteristic geometric elements of the higher order root locus plots on the other hand. These correlations in connection with some of the geometric peculiarities, mentioned in Chapter VI.1, may be used successfully for synthesis purposes to improve systematically the behaviour of a multi-loop control system or of a mechanical or electrical network in a prescribed sense.

In linear multi-loop control systems, the higher order root locus technique may provide a compromise between classical design technique which , very often , turns out to be rather inflexible and which doesn't guarantee enough freedom in selecting system parameters, and between modern design methods which may introduce too much freedom in selecting or computing system parameters (compare Chapter VII).

The amount of additional freedom in system design by applying higher order root locus techniques compared to classical root locus techniques may be illustrated by the following simple example : Assume, the open-loop transfer function of a linear control system has m zeros, at least m + 3 poles, and a characteristic polynomial which is linear in a feedback factor k. Then it is impossible to stabilize the system for sufficiently large values of k by changing a coefficient of the characteristic polynomial.

On the other hand , if various (at least two) parameters of a control system are changed simultaneously and proportionally to each other or to another parameter k , then the caracteristic polynomial may become nonlinear in k , and there may exist a chance to stabilize the system for large values of k by changing coefficients of the characteristic polynomial, despite the fact that the number of poles of the system transfer function matrix exceeds the number of its zeros or sinks by more than two. This is shown in example 6.2 .

The residual examples demonstrate how to stabilize an unstable system for small values of $|k|$, how to change selectively the asymptote points, the asymptote angles, the structure of the asymptote pattern or the sinks and return points of the system in order to stabilize the system for arbitrary values of k , how to detect critical parameter dependencies, or how to detect or improve certain robustness features of the system. In all of these situations, the structure of the exponent diagram gives strong hints how to achieve these aims without affecting other features of the root locus plot. This procedure can even be used to improve the behaviour of nonlinear systems, as will be shown in connection with the "describing root locus techniques" in a later paper.

Example 6.1 :

Figures 5.1b and 5.1c of Chapter V show that the system of example 5.1 is unstable both for small and for large positive values of the parameter k and for (a,b) = (2,-3).

Question : Is it possible by selecting suitable values of (a,b) to stabilize the system

(i) for small positiv values of k ?

(ii) for all values of k (robust system) ?

Answer :

As is shown in Table 5.a.1 , the coefficient b affects the segments I and III of the exponent diagram for $a \neq o$ and in addition segment VII for $a = o$.
As a consequence, the angles of departure of root locus branches from poles s_j^o of the open loop transfer function of the system may be influenced by the parameter b (for $b \notin S := \{o,-1,-2\}$).
For $b \in S$, the angle of departure of a root locus branch from point s_j^o depends on the value of the coefficient a .
Moreover, the coefficient b may influence both, the number of root locus branches meeting the points s_j^o for real values $k_{lo}^j \neq o$ and the real roots k_{lo}^j of $R_o(k)$.
The number and the position of return points are affected by the coefficient b, too.

The coefficient a affects the supporting polynomial related to segment VIII (i.e. the asymptote angles of the root locus plot) and the roots k_{lo}^j of the polynomial R_{lo} (i.e. the number of root locus branches meeting the points s_j^o for real values $k_{lo}^j \neq o$), the values k_{lo}^j , the return points and (for $b = o$) the angles of departure from points s_j^o).

Answer to question (i) :

Let (a,b) = (-2,+3) and (a,b) = (2,3).
In both cases the system gets stable for small positiv values of k (compare Table 5.c.1 and Rule 1o of example 5.1).
The root locus plots are easily computed in line to Example 5.1, Chapter V . They are shown in Figures 6.1.a, 6.1.b, 6.1.c and 6.1.d for $k \geq o$.

Answer to question (ii) :

In agreement with Rule 14 (Chapter IV), the asymptotes constitute a Butterworth Configuration for arbitrary non zero values of coefficient a ($a \neq o$). Therefore, at least one of the three root locus branches enters the unstable region of the complex plane for large values of k. Only for $a = o$ it is possible to keep the two asymptotes away from the unstable region for arbitrary values of k (k either positive or negative). The case $a = o$ is easily treated by classical root locus technique. (comp. Fig. 6.1e for $k \geq o$).

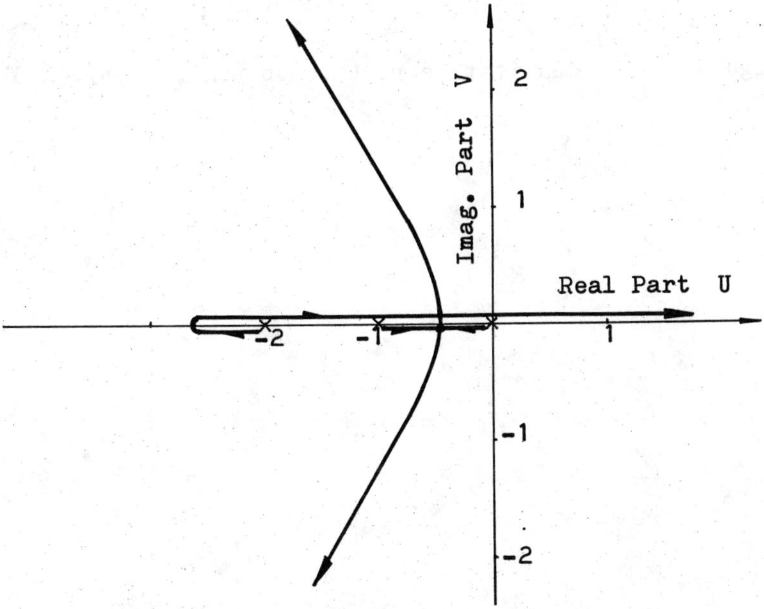

Figure 6.1.a: Qualitative root locus plot,(a,b)=(-2,3) .

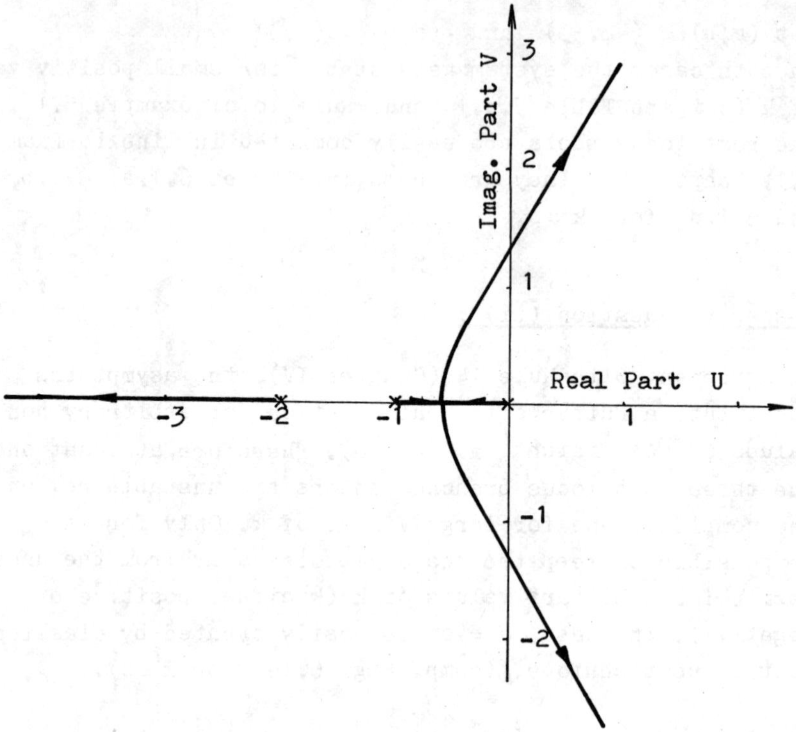

Figure 6.1.b: Qualitative root locus plot, (a,b)=(2,3) .

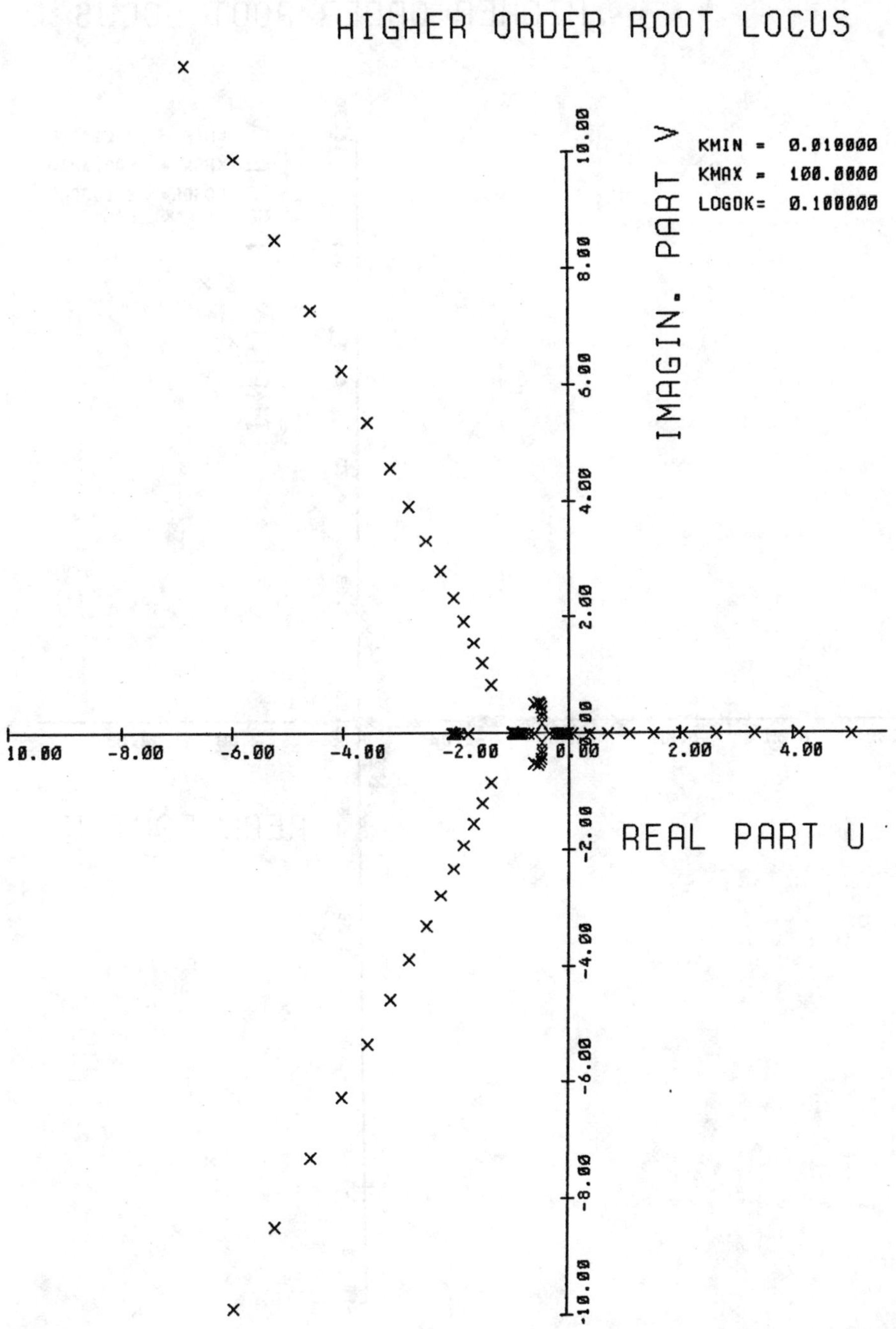

Figure 6.1.c: Computer root locus plot, (a,b)=(-2,3).

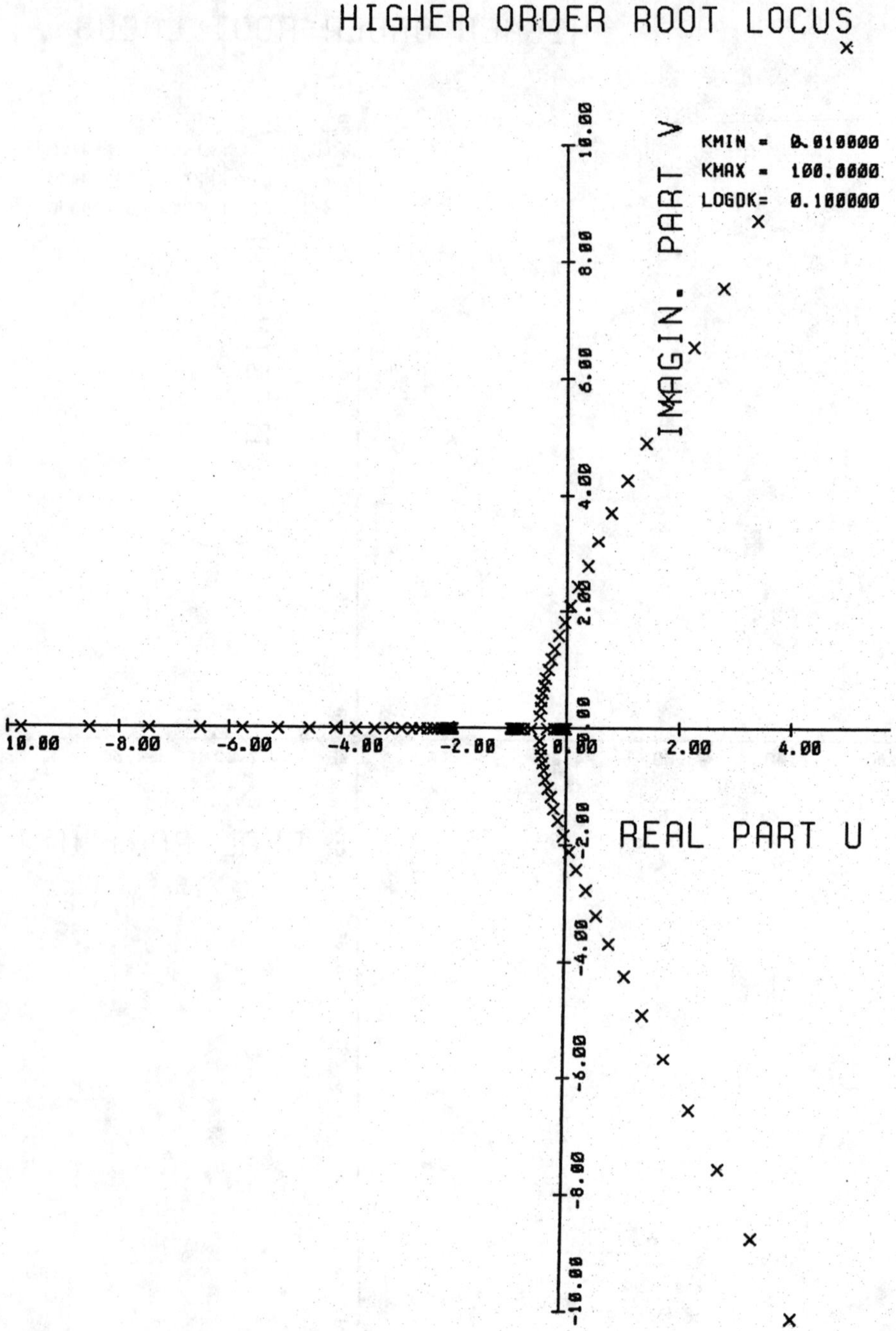

Figure 6.1.d: Computer root locus plot, (a,b)=(2,3).

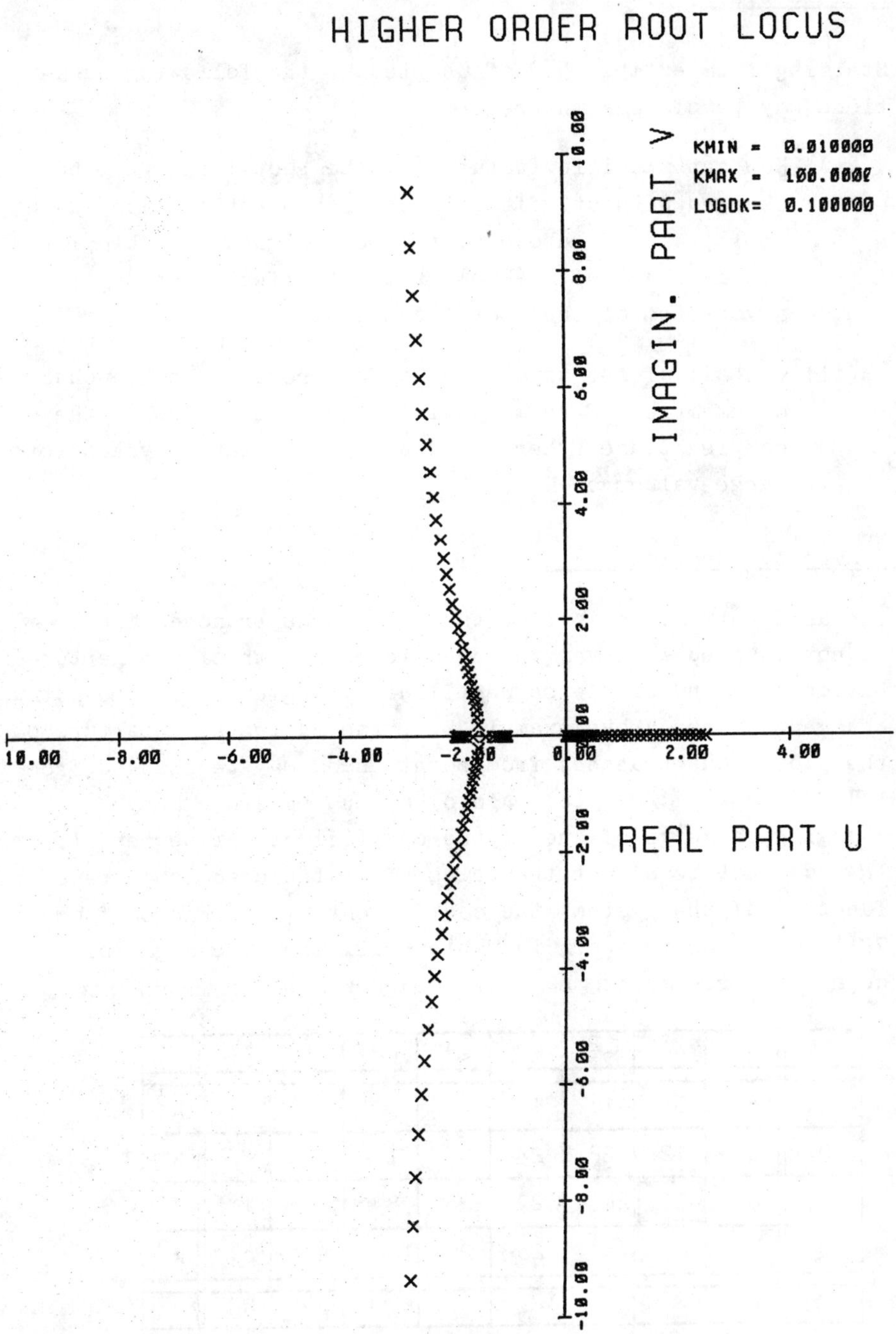

Figure 6.1.e: Computer root locus plot (a,b)=(o,-3).

Example 6.2 :

Starting from example 5.2 of Chapter V, the following questions may be of some interest :

(i) How can specific features of the root locus plot be changed without affecting other properties. For example, can the above system be designed unstable for small values of k (k>o) without affecting the asymptotes of the root locus plot ?

(ii) Can all of the asymptotes of the root locus branches of example 5.2 be moved into the stable part of the complex plane (thereby ensuring a stable system for large values of k) ?

Answer to question (i) :

The asymptotic behaviour of the root locus branches for $k \to \infty$ is not changed whenever the coefficients both of the perturbation line and of the segment line of segment VIII are not changed. In the above example 5.2, the angles of departure of the root locus branches from point s=o may be changed from $+\pi/2$ and $-\pi/2$ to $\varphi(o,o)=o$ and $\varphi(o,o)=\pi$ by changing the sign of one of the coefficients of segment I. In order not to affect the poles s_j^o of the open loop transfer function of the system, the coefficient a_1^1 is changed from 72 to -72 as is shown in Table 6a.2. Then the angle of departure from point s=-4 is changed from o to π ,too.

	s^0	s^1	s^2	s^3	s^4	s^5	s^6	s^7	
Q_3	o	o	1	1					k^3
Q_2	o	12	28	23	8	1			k^2
Q_1	o	-72	246	329	22o	78	14	1	k^1
Q_0	o	o	o	2o	29	1o	1	o	k^0
	R_0	R_1	R_2	R_3	R_4	R_5	R_6	R_7	

Table 6.a.2: Coefficients of Polynomials $Q_j(s)$ and $R_i(k)$.

Q_i	i	$s-s_1^i$	$s-s_2^i$	$s-s_3^i$	$s-s_4^i$	$s-s_5^i$	$s-s_6^i$	$s-s_7^i$	
Q_3	3	s+o	s+o	s+1					k^3
Q_2	2	s+o	s+1	s+2	s+2	s+3			k^2
Q_1	1	s+o	s-o,22	s+4,9	s+1 -i.1,75	s+1 +i.1,75	s+3,6 -i.1,75	s+3,6 +i.1,75	k^1
Q_0	o	s+o	s+o	s+o	s+1	s+4	s+5		k^0
		R_0	R_1	R_2	R_3	R_4	R_5	R_6	

Table 6.b.2: Roots of polynomials $Q_i(s)$

	s^0	s^1	s^2	s^3	s^4	s^5	s^6	s^7	
Q_2			k+14 -i.7,o7	k+11,5 -i.14					$k-k_j^2$
Q_1		k-6	k+14 +i.7,o7	k+11,5 +i.14	k+27,4	k+77,9			$k-k_j^1$
Q_0		k+o	k+o	k+o,o6	k+o,13	k+o,128	k+o,o7	k+o	$k-k_j^0$
j	o	1	2	3	4	5	6	7	
R_j	R_0	R_1	R_2	R_3	R_4	R_5	R_6	R_7	

Table 6.c.2: Roots of polynomials $R_j(k)$

As a consequence, only the roots of the polynomials $Q_1(s)$ and $R_1(k)$ are changed as indicated by Tables 6.b.2 and 6.c.2 .

The point $s = o$ is met by a (real) root locus branch for $k_{1o}^{j} = o$ and for $k_{1o}^{2} = +6$. As a consequence, there exists a real return point for $o < k < 6$ (compare Figures 6.2.a and 6.2.b for $k \geq o$). In addition, point $s = -1$ is no longer a degenerated root locus branch.

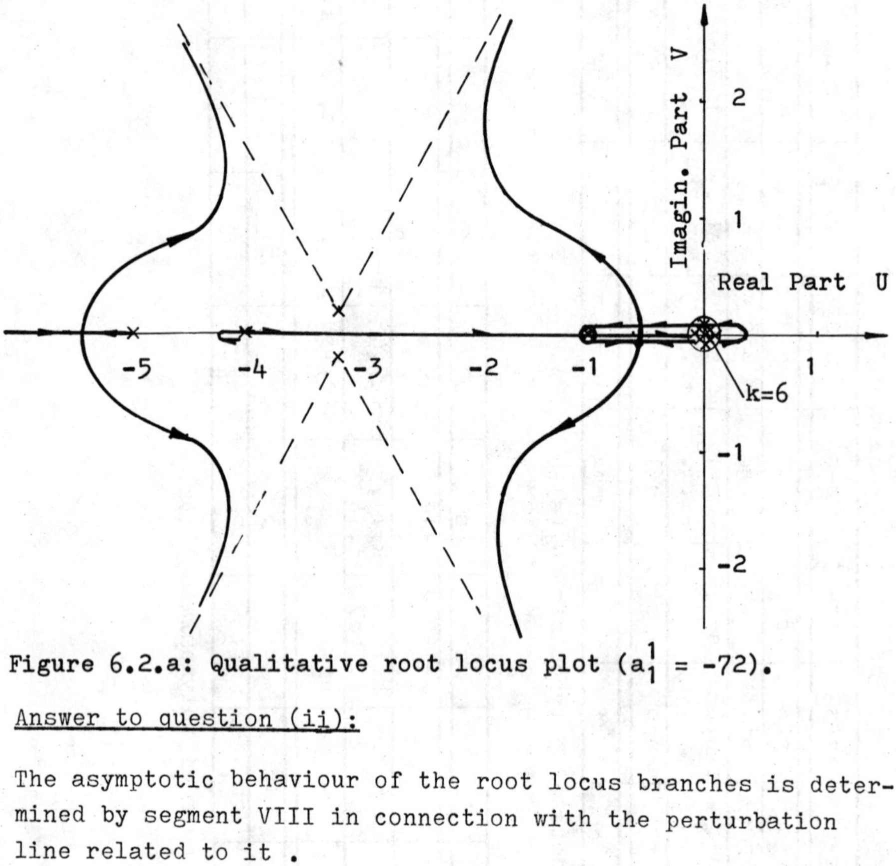

Figure 6.2.a: Qualitative root locus plot ($a_1^1 = -72$).

Answer to question (ii):

The asymptotic behaviour of the root locus branches is determined by segment VIII in connection with the perturbation line related to it .

Let the coefficient a_3^3 of the characteristic polynomial be changed from $a_3^3 = 1$ to $a_3^3 = o{,}2$ (compare Table 6.d.2).

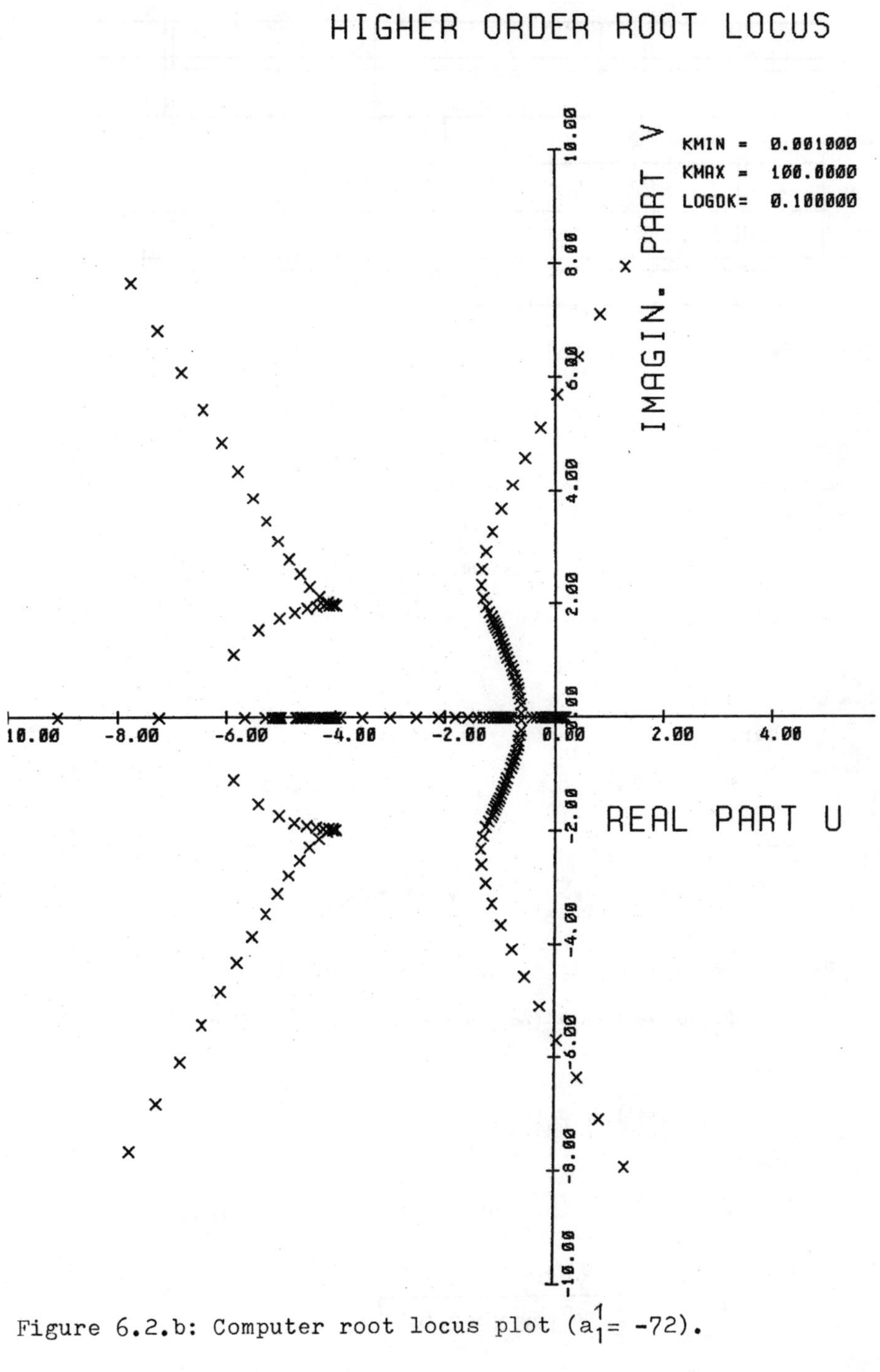

Figure 6.2.b: Computer root locus plot ($a_1^1 = -72$).

	s^0	s^1	s^2	s^3	s^4	s^5	s^6	s^7	
Q_3	o	o	1	o,2					k^3
Q_2	o	12	28	23	8	1			k^2
Q_1	o	72	246	329	22o	78	14	1	k^1
Q_0	o	o	o	2o	29	1o	1	o	k^0
	R_0	R_1	R_2	R_3	R_4	R_5	R_6	R_7	

Table 6.d.2 : Polynomial coefficients of $Q_j(s)$ and $R_i(k)$

As a consequence :

a. the sink s_3^3 of the system is changed from $s_3^3 = -1$ to $s_3^3 = -5$,

b. the roots of the polynomial $R_3(k)$ are changed,

c. the supporting polynomial of segment VIII has the form

$$\gamma^4 + \gamma^2 + 0,2 \quad ,$$

where

$$\gamma_1 = + i.\ 0,85 \quad , \quad \gamma_2 = - i.\ 0,85$$

and

$$\gamma_3 = + i.\ 0,53 \quad , \quad \gamma_4 = - i.\ 0,53$$

d. the asymptote angles take the form

$$\varphi_1(\infty,\infty) = \varphi_2(\infty,\infty) = \frac{\pi}{2} \quad , \quad \text{for } k > 0$$

and

$$\varphi_3(\infty,\infty) = \varphi_4(\infty,\infty) = -\frac{\pi}{2} \quad , \quad \text{for } k > 0 \quad ,$$

and

e. the asymptote points are computed from relation

$$s_0 = \frac{14.\ \gamma^6 + 8.\ \gamma^4 + \gamma^2}{7.\ \gamma^6 + 5.\ \gamma^4 + 0,6.\gamma^2}$$

or

$$s_{o1} = -3{,}95 \quad , \quad s_{o2} = -3{,}95$$

$$s_{o3} = -0{,}565 \quad , \quad s_{o4} = -0{,}565 \quad .$$

The root locus plot is drawn in Figure 6.2.c for $k > 0$ (this system is robust with respect to parameter variations of k). On the other hand, it is not robust with respect to variations of the coefficient a_3^3 (compare Figures 5.2c, 6.2c and 6.2d for values $a_3^3 = 1$, $a_3^3 = 0{,}2$ and $a_3^3 = 0{,}1$, respectively, and for $k > 0$).

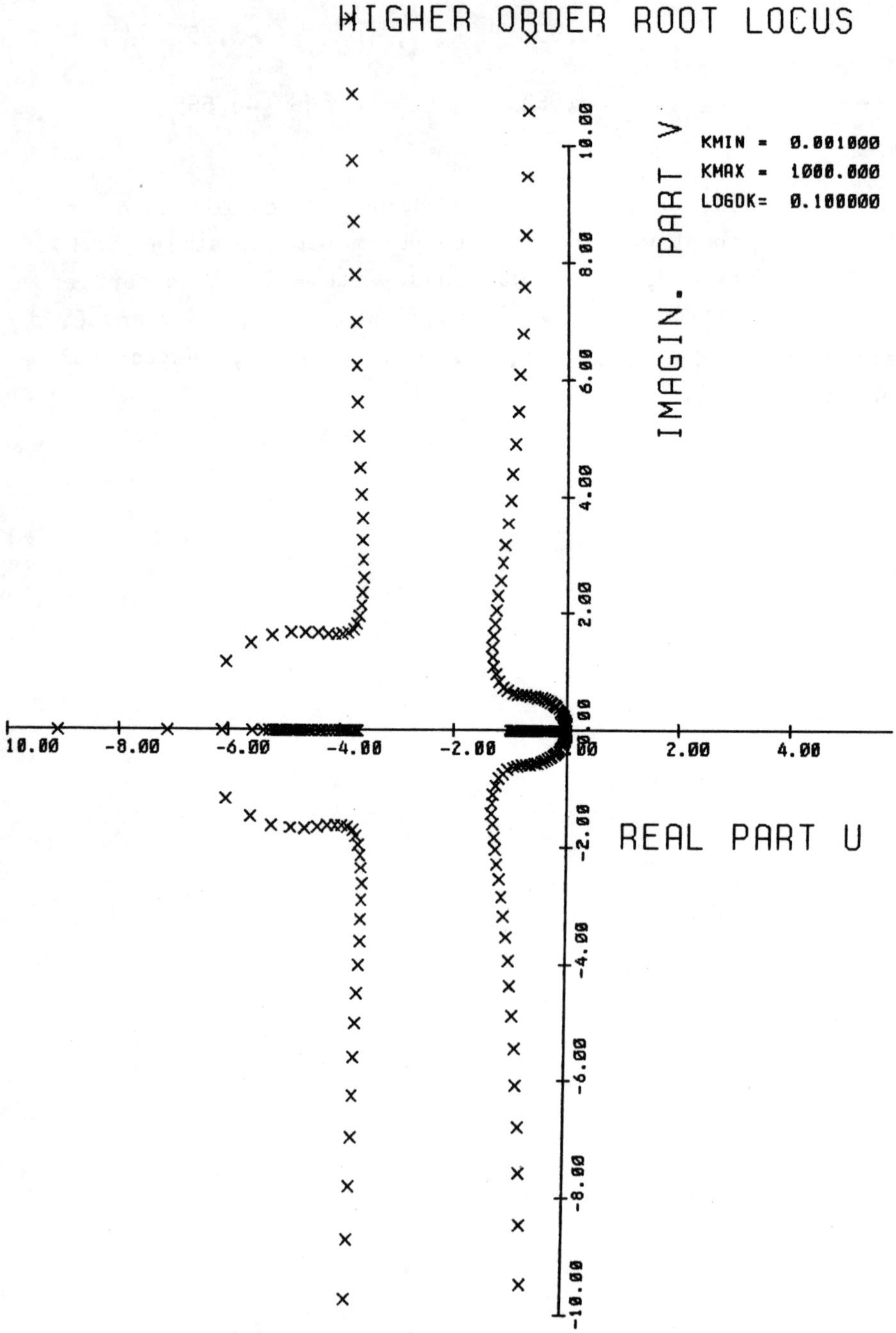

Figure 6.2.c: Computer root locus plot (a_3^3= o,2).

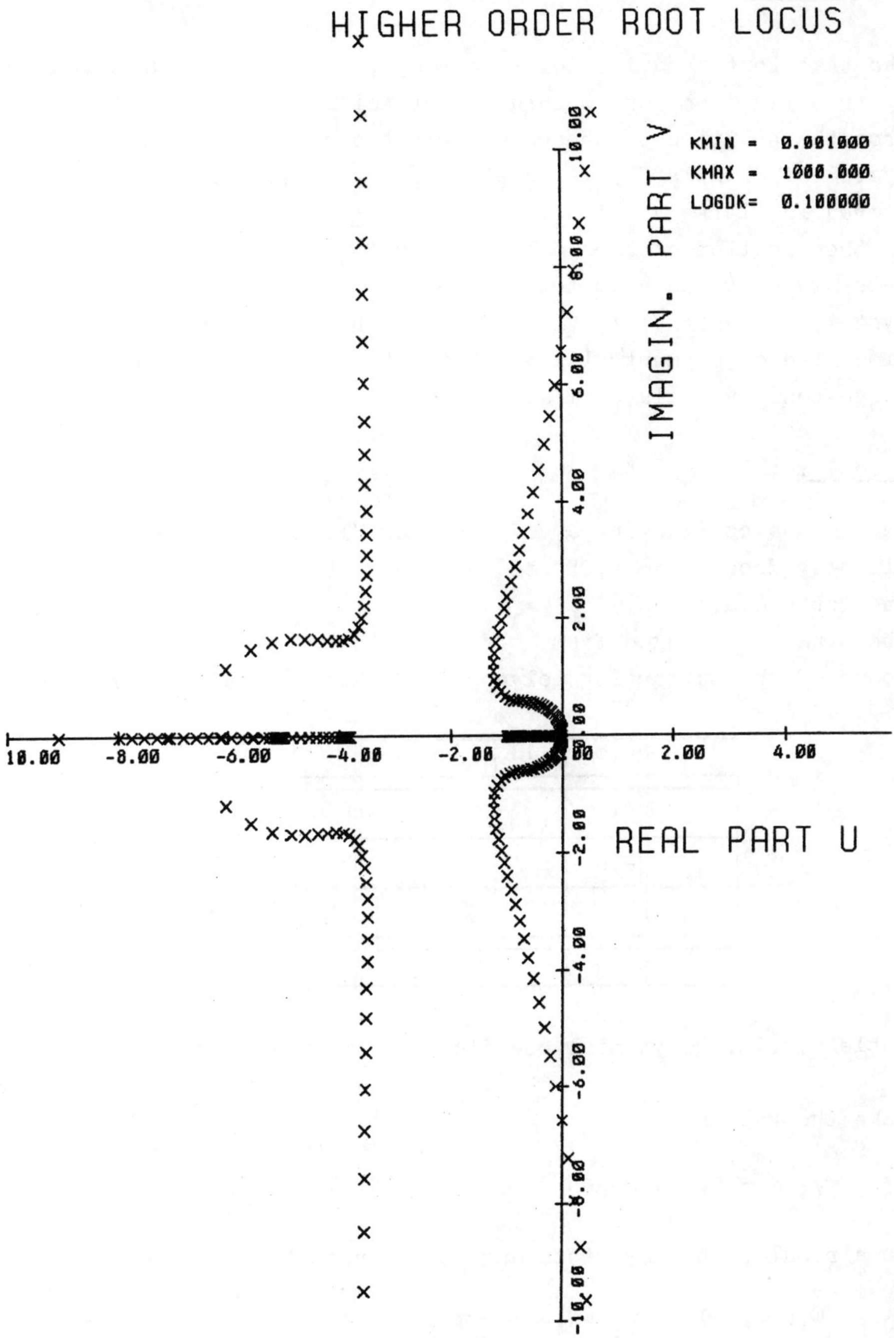

Figure 6.2.d: Computer root locus plot (a_3^3= o,1).

Example 6.3 :

The transient behaviour of linear systems which are described by transfer functions with dominant poles may be estimated from the position of these poles in the complex plane. The system of example 5.3 in Chapter V, for instance, is over-damped for large values of k, $k > o$.
As the position of closed loop poles for large values of k is determined by the asymptotes as well as by the sinks of the system, segments VIII, VII and V together with the perturbation line of segment VIII determine the closed loop poles for large positive values of k.

Case 1 :

The common coefficient a_1^2 of segments VII and VIII is changed slightly from $a_1^2 = 0,25$ to $a_1^2 = 0,2$ (compare example 5.3 and Table 6.a.1)
The sink has changed from $s_1^2 = -4$ to $s_1^2 = -5$, and the roots of the supporting polynomial related to segment VIII

	s^0	s^1	s^2	s^3	
Q_2	1	o,2			k^2
Q_1	-16	o	1		k^1
Q_0	o	2	3	1	k^0
	R_0	R_1	R_2	R_3	

Table 6.a.1 : Polynomial coefficients of $Q_j(s)$ and $R_i(k)$.

take the values

$$\gamma_1 = -0,723 \quad \text{and} \quad \gamma_2 = -0,278 \quad .$$

As a result, the asymptote angles are not changed, i.e.

$$\varphi_1(\infty,\infty) = \varphi_2(\infty,\infty) = \pi \quad .$$

On the other hand, in line to Rule 13 we have

$$\varphi(-5,\infty) = 0 \quad , \quad \text{for} \quad k \geq 0 \quad .$$

This follows from

$$Q_2(s') = 0{,}2.s' + \ldots \qquad \text{and}$$

$$Q_1(s') = 9 + \ldots$$

by replacing s in $Q_2(s)$ and $Q_1(s)$ by $s'-5$ and by applying (4.26b).
The change of the angle of arrival from

$$\varphi(-4,\infty) = \pi \qquad \text{to} \qquad \varphi(-5,\infty) = 0$$

implies, that the root locus plot has in addition to the plot of Figure 5.3c another curved (complex conjugate) branch.
As a consequence, the transient behaviour of the modified system shows an increasing tendency for oscillation for values $7 < k < 9o$ compared to example 5.3 .

The return points are computed from the relations

$$R(P, \partial P/\partial k) = \det \begin{pmatrix} Q_2 & Q_1 & Q_0 \\ 0 & 2.Q_2 & Q_1 \\ 2.Q_2 & Q_1 & 0 \end{pmatrix} = Q_2.(Q_1.Q_1 - 4.Q_0.Q_2)$$

or

$$R = 0{,}2.(s+5).(\ (s+4)^2.(s-4)^2 - 0{,}2.4.s.(s+1)(s+2)(s+5))$$

$$= 0{,}2.(s+5).((s^2-16)(s^2-16) - 0{,}8.(s^3+3s^2+2.s).(s+5))$$

$$= 0{,}2.(s+5).(0{,}2.s^4-6{,}4s^3-45{,}6s^2-8s+256) \ .$$

They are changed to

$$s = 2{,}04 \qquad \text{for} \quad k > 0$$

and

$$s = 38 \qquad \text{for} \quad k < 0 \quad .$$

The root locus plots are shown in Figures 6.3.a and 6.3.b for $k \geq 0$.

<u>Case 2 :</u>

The transient behaviour of the system may be further improved by changing the asymptote angles. This can be achieved by changing again the common coefficient a_1^2 of the segments VII and VIII from

$$a_1^2 = 0,2 \quad \text{to} \quad a_1^2 = 2,5 \quad .$$

	s^0	s^1	s^2	s^3	
Q_2	1	2,5			k^2
Q_1	-16	o	1		k^1
Q_0	o	2	3	1	k^0
	R_0	R_1	R_2	R_3	

Table 6.3.a.2 : Polynomial coefficients of $Q_j(s)$ and $R_i(k)$.

Then the sink is changed to $s_1^2 = - 0,4$.

The corresponding angle of arrival at $s = - 0,4$ for $k = \infty$ is

$$\varphi(-0,4;\infty) = \pi$$

in agreement with Rule 13 .

The supporting polynomial of segment VIII

$$\gamma^2 + \gamma + 2,5 = 0$$

has the roots

$$\gamma_1 = - \tfrac{1}{2}\cdot(1+3.i) \quad \text{and} \quad \gamma_2 = - \tfrac{1}{2}\cdot(1-3.i) \quad .$$

It provides asymptote angles

$$\varphi_1(\infty,\infty) = 108,5^o \quad \text{and}$$

$$\varphi_2(\infty,\infty) = 251,5^o \quad ,$$

and asymptote points

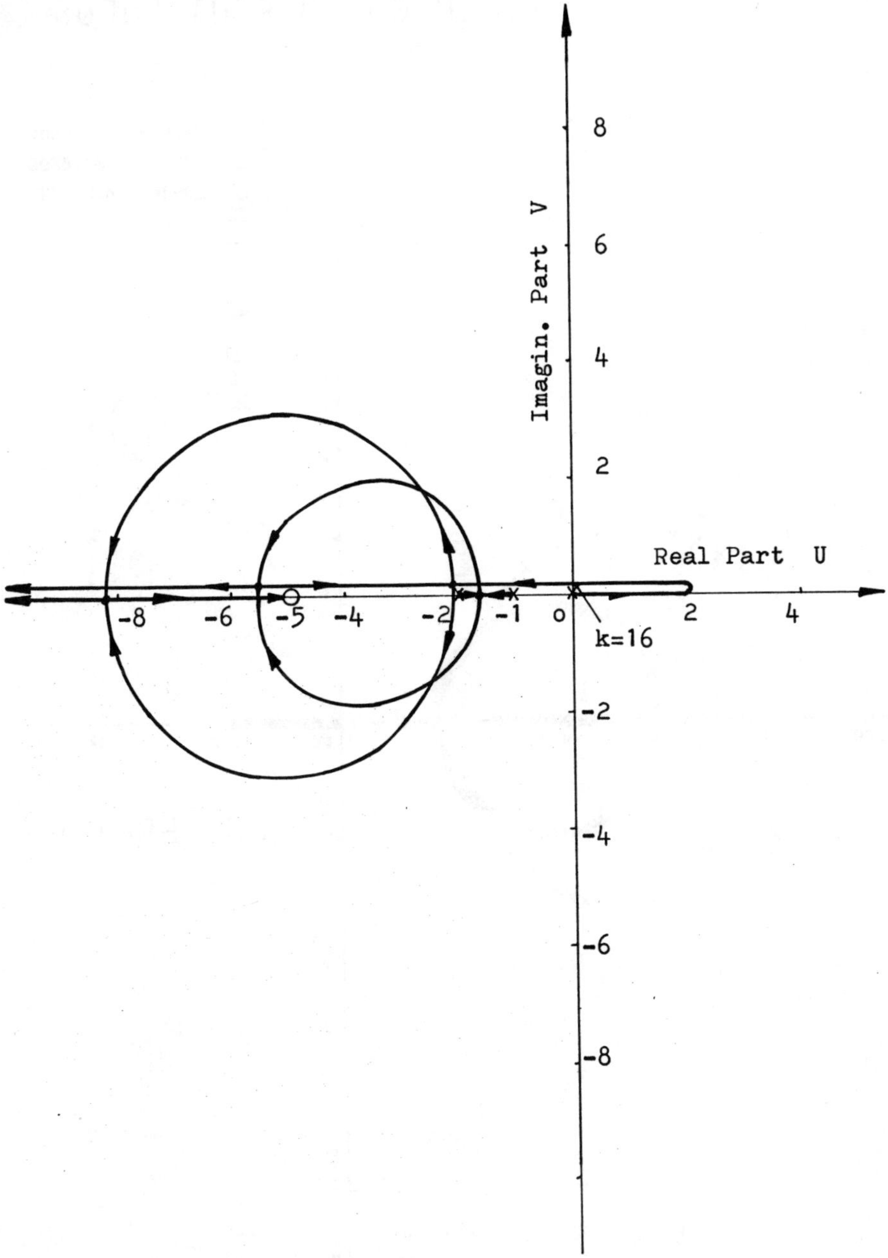

Figure 6.3a: Qualitative root locus plot (a_1^2 = o,2) .

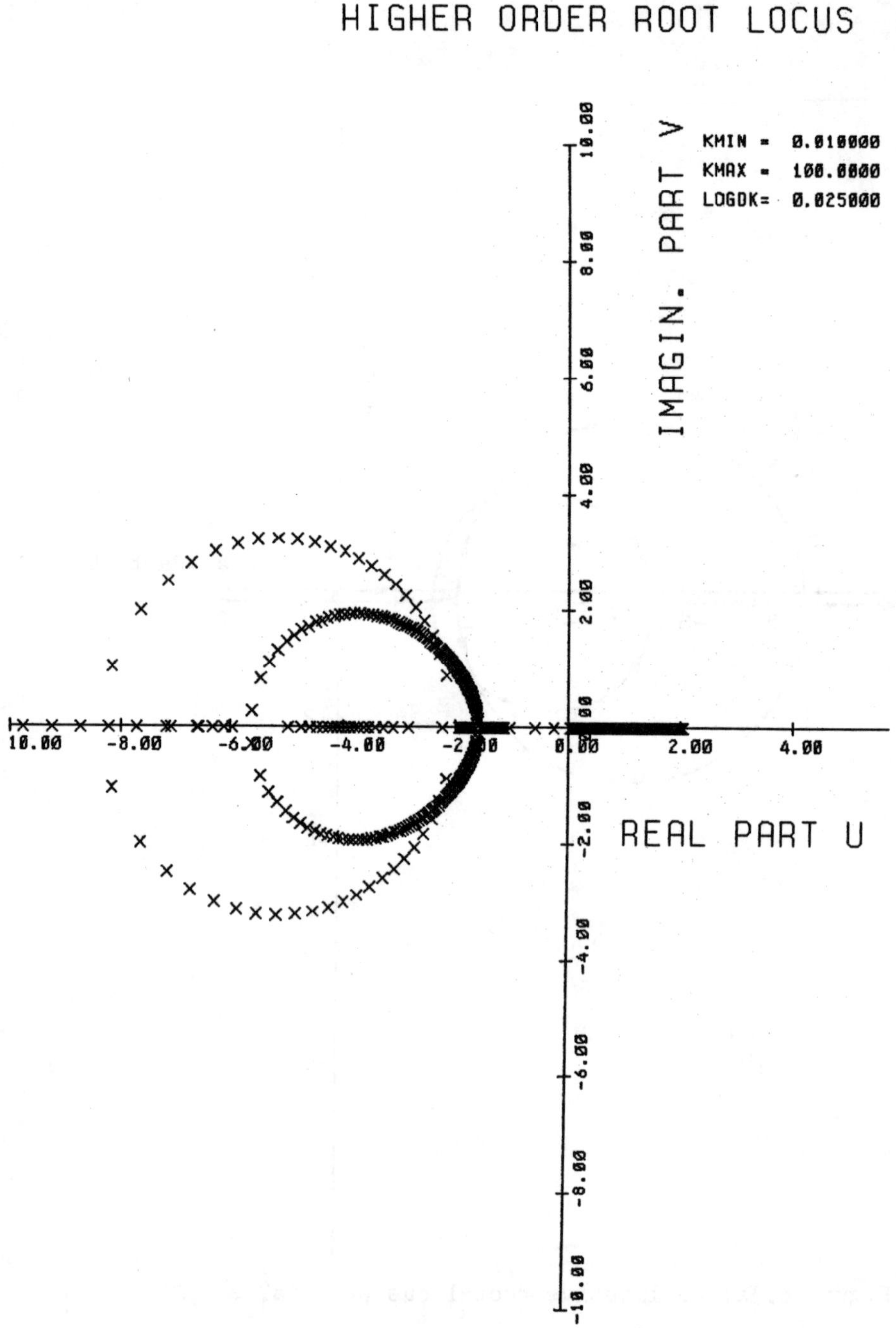

Figure 6.3.b: Computer root locus plot (a_1^2= o,2).

$$s_{o_j} = - \frac{3 \cdot \gamma_j^2 + 1}{3 \cdot \gamma_j^2 + 2 \cdot \gamma_j + 2,5} \quad ; \quad j = 1,2$$

or

$$s_{o_1} = - 1,3 + 0,55 \cdot i \quad , \quad s_{o_2} = - 1,3 - 0,55 \cdot i \quad .$$

The return points take the values

$$s_1 = 1,394 \quad \text{for} \quad k > 0$$

and

$$s_2 = - 2,585 \quad \text{for} \quad k < 0 \quad .$$

The root locus plots are drawn in Figures 6.3.c and 6.3.d for $k > 0$.

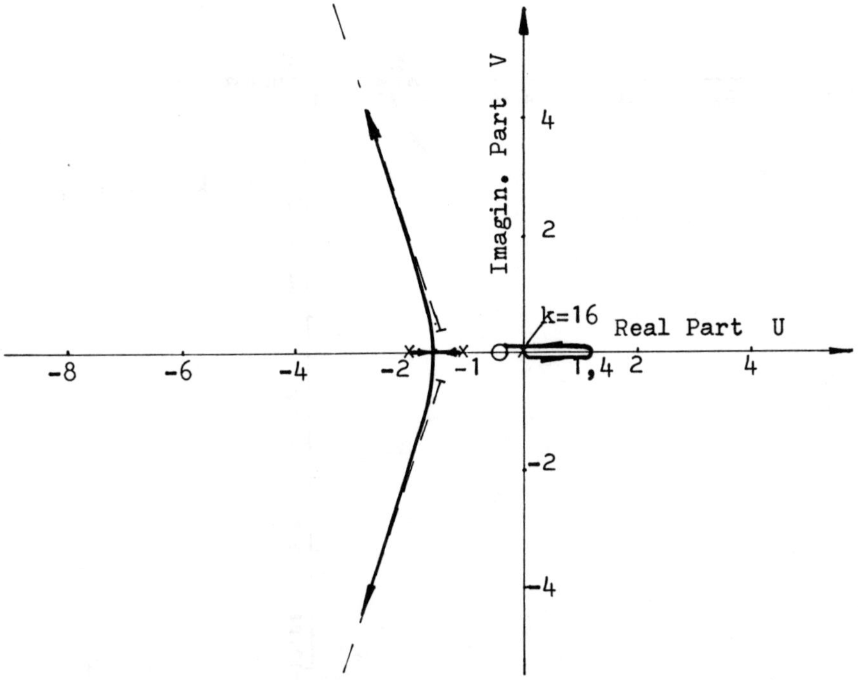

Figure 6.3.c: Qualitative root locus plot ($a_1^2 = 2,5$) .

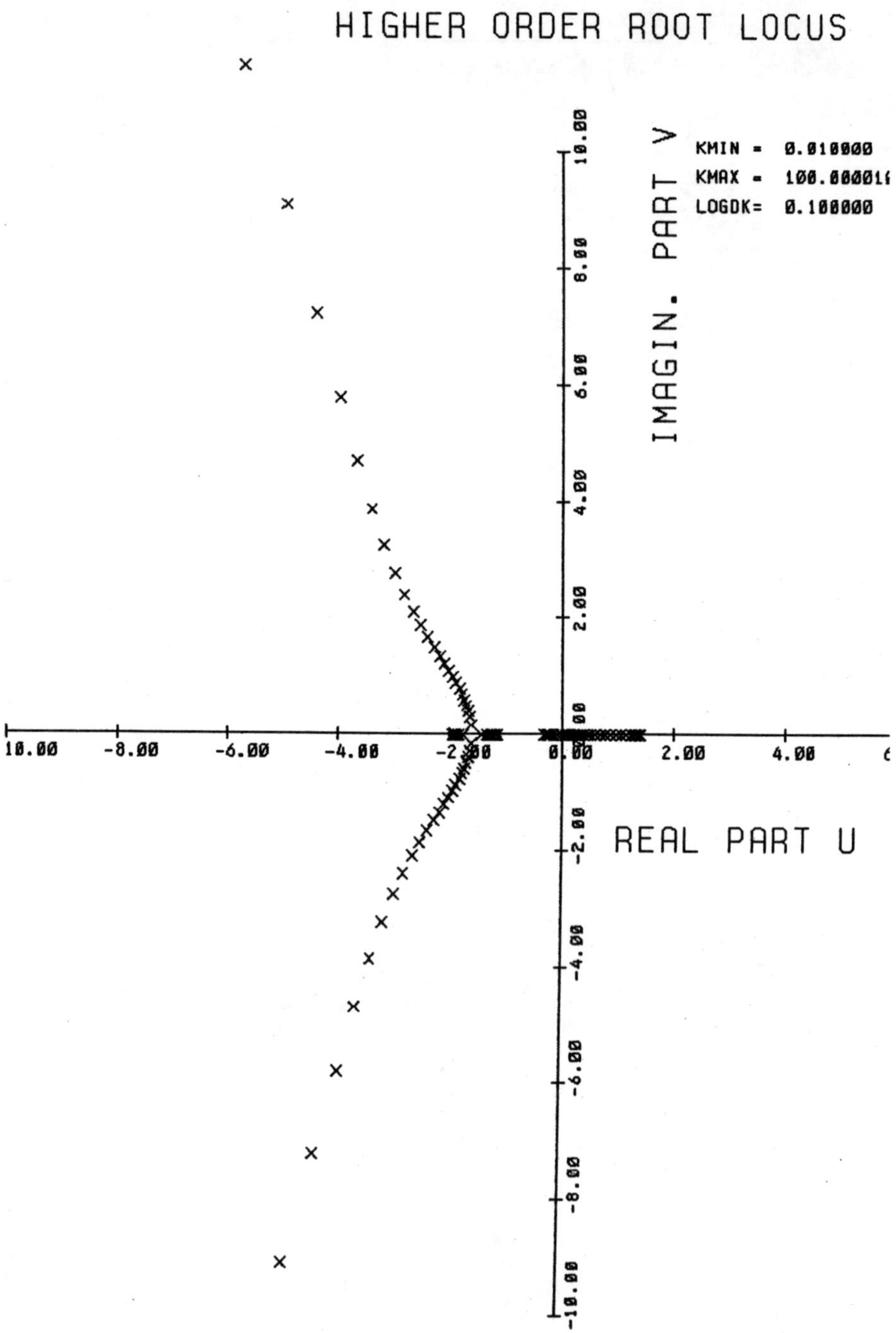

Figure 6.3.d: Computer root locus plot ($a_1^2 = 2{,}5$).

Example 6.4 : (Cross coupling of airplane)

The sensitivity problem of linear control systems is often treated in terms of the parameter dependence of the poles of the closed loop transfer function. Higher order root locus technique provides a useful tool both for local and global eigenvalue sensitivity analysis. The exponent diagram yields much information concerning the dependence of specific eigenvalues on selected coefficients of the characteristic polynomial.

Let the coefficient $a_o^2 = 0{,}817$ of example 5.4 be changed to $a_o^2 = 0{,}8$ and to $a_o^2 = 0{,}86$, respectively.

This coefficient affects the asymptotes as well as the roots k_{1o}^j of $R_{1o}(k)$ and the return points of the system.

In the <u>first</u> case ($a_o^2 = 0{,}8$) , $R_o(k)$ has two real roots $k_o^1 = + 7{,}08$ and $k_o^2 = + 9{,}52$, the point $s = 0$ is met twice by a root locus branch, and there exist a real unstable return point (for $s > 0$).

In the <u>second</u> case $R_o(k)$ has only complex roots. No branch meets the point $s = 0$,and there are no real return points for $k > 0$ (compare Figures 6.4.a and 6.4.b).

This example shows, that the exponent diagram may be used effectively to determine a critical parameter dependence of interesting system coefficients.

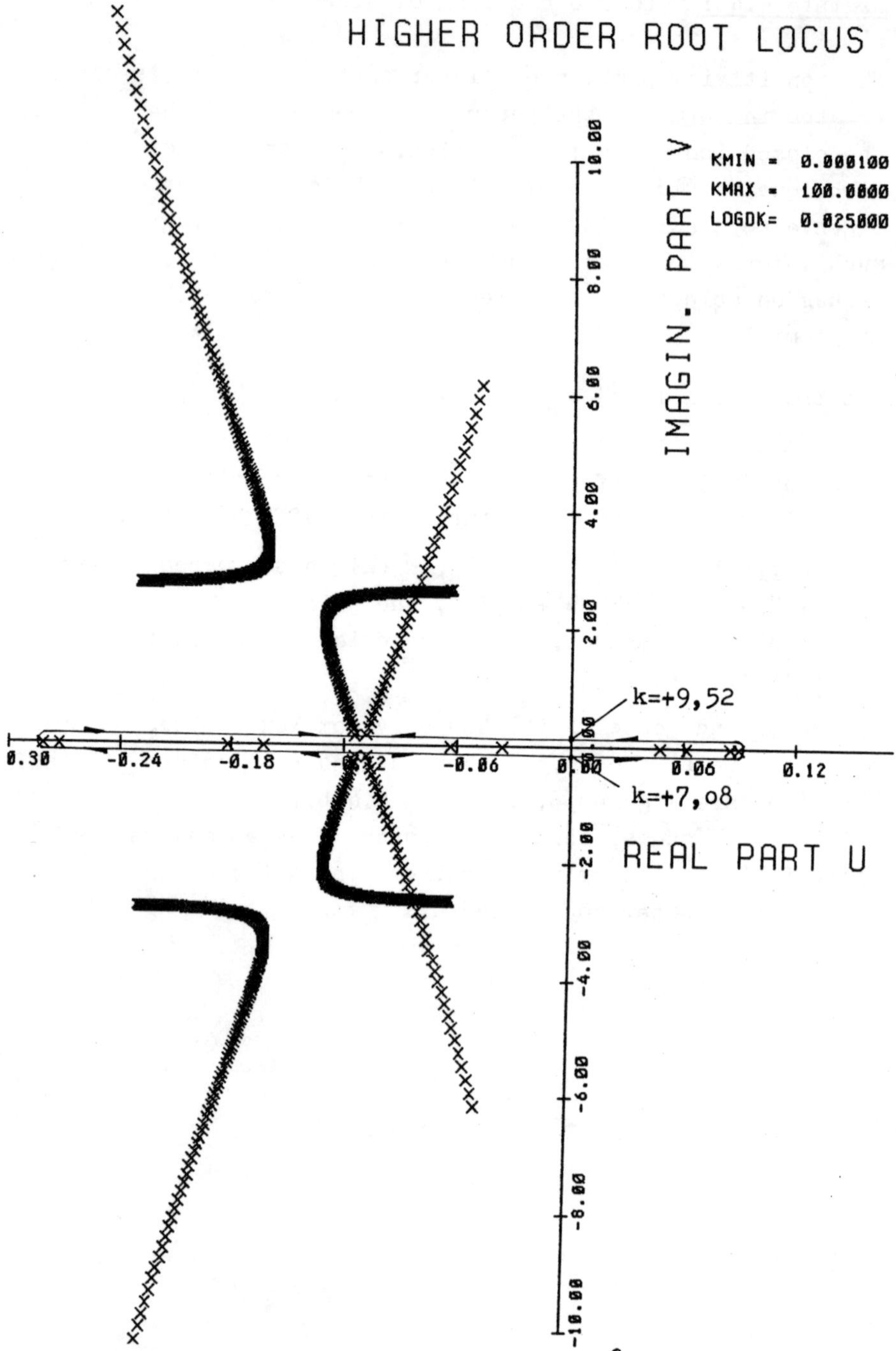

Figure 6.4.a: Computer root locus plot (a_o^2 = o,8).

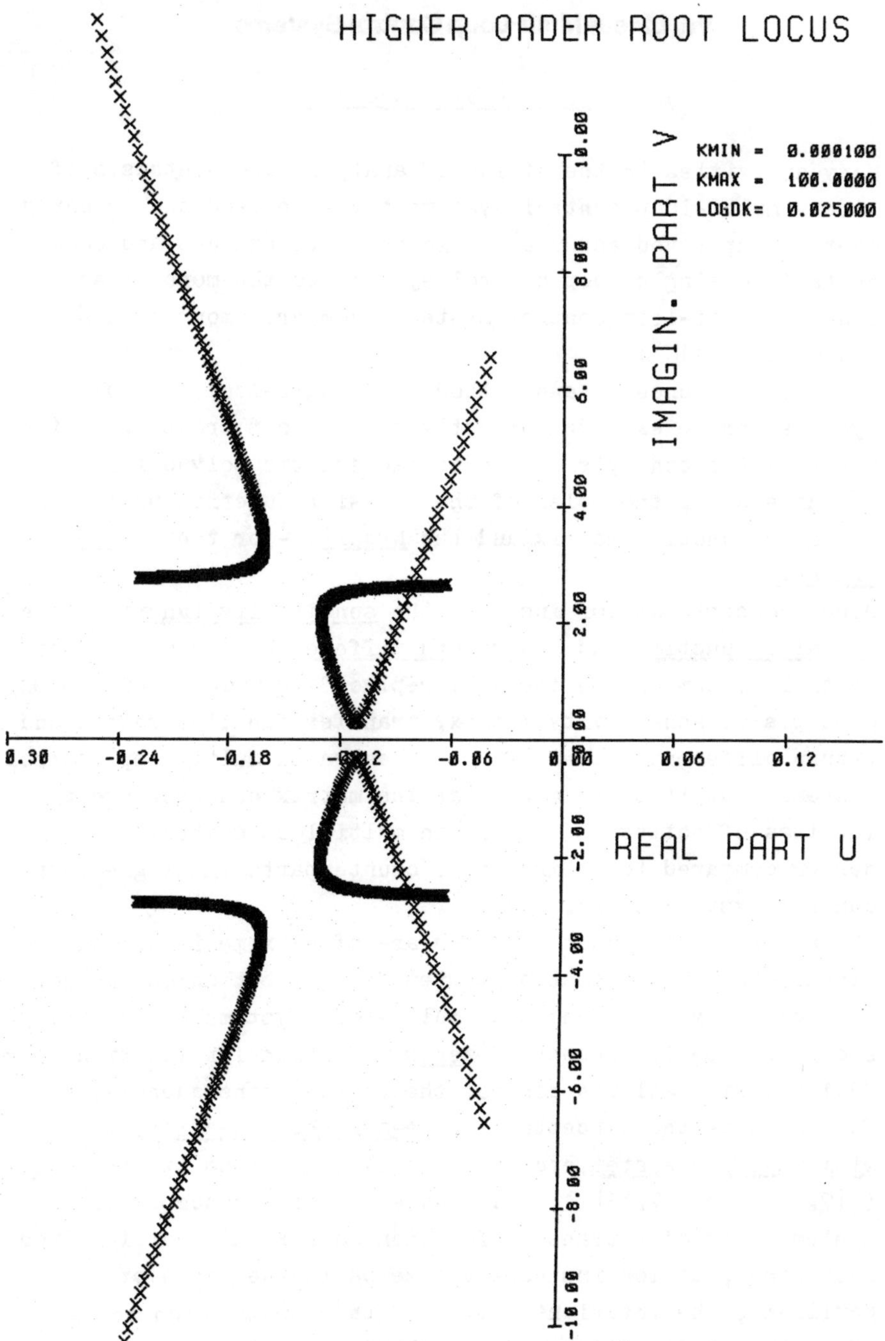

Figure 6.4.b: Computer root locus plot ($a_o^2 = o{,}86$).

VII. Linear Multi-Loop Control Systems.

1. General concepts.

Many activities in the fields of analysis and synthesis of linear multi-loop control systems try to extend and to carry over the approved and useful synthesis techniques and concepts from single-loop control systems to the more general case of multi-loop control systems (compare among others [7.1] to [7.11]).

Some of the concepts and methods of single-loop control systems can be extended directly and without trouble, as for instance the concepts of the characteristic polynomial of a linear system, the poles of the transfer function or of the transfer function matrix and the Nyquist - or the Circle criterions.

Other concepts as for instance the sensitivity function, the transfer function and the return difference of a single-loop control system can be formally replaced without trouble by the concepts of sensitivity matrix, transfer function matrix and return difference matrix for the design of multi-loop control systems. But it turns out, that the matrix concepts are by far less effective and useful in multi-loop control system design compared to their scalar counterparts in single-loop control system design.

There are other concepts, which are of extreme importance in linear single-loop control system design, but which can not even uniquely be extended to multi-loop systems as for instance the basic concept of zeros of a transfer function [7.12] , with all its control theoretical consequences, as for instance the concepts of pole-zero-compensation, or minimum-phase-system,etc. As is well known from literature ([7.13] to [7.16]), there exist various concepts of system zeros of a transfer function matrix which differ from each other, as for instance the zeros in the sense of McMillan , the invariant zeros or the transmission zeros of a linear multi-loop system. These concepts are necessary refinements of the zeros of a single loop control system.

Independent of these various concepts of multi-loop system zeros, the general concept of a <u>sink</u> of a multi-loop system has been introduced by the author in connection with the higher order root locus technique ([7.17],[7.18]). This will be discussed in what follows.

In the domain of multi-loop control systems, various design methods are known. Among those, the classical "trial and error" techniques and the modern more formal procedures are typical representatives.

The classical design techniques are usually based upon the variation of a scalar parameter (e.g. a single scalar feedback factor k) for a single control loop and by repeating this procedure, step by step, for each control loop. This can be done by a repeated application of classical root locus techniques. Because of the poor structure of classical root locus technique (compare Chapter VI), this method is rather narrow banded and rigid in connection with multi-loop control systems. It doesn't provide enough freedom in choosing system parameters.

Modern more formal design procedures as for instance the optimal controller design on the basis of the Riccati Equation or the arbitrary state feedback pole placement design often provide too much freedom in choosing optimal control parameters. In the first case, the choice of valuable elements of the weighting matrices Q and R of the optimization criterion (compare Chapter VIII) may be troublesome and problematic. In the second case, the nonlinear algebraic equations in the system parameters may have much more solutions than unknown parameters.

A compromise between the above two extremal approaches, which may be useful in specific situations of practice, may be formulated as follows :

Given a control plant of the form

$$\dot{x} = A.x + B.u \quad , \quad y = C.x \qquad (7.1.a)$$

with transfer function matrix of the form

$$P(s) = C.(s.I - A)^{-1}.B = \frac{C.adj^T(s.I-A).B}{det(s.I-A)} ,$$

where $x \in \mathbb{R}^n$ and $u,y \in \mathbb{R}^p$. (7.1.b)

Using a state feedback matrix K , the closed-loop system transfer function matrix takes the form

$$G_c(s) = C.(s.I_n - A + B.K)^{-1}.B \tag{7.2.a}$$

or

$$G_c(s) = \frac{C.adj^T(s.I-A+B.K).B}{det(s.I - A + B.K)} , \tag{7.2.b}$$

with related characteristic polynomial

$$\Psi_c(s;k) = det\ (s.I_n - A + B.K). \tag{7.2.c}$$

Application of classical root locus techniques to (7.2.c) with K as (n.n) - parameter vector leads to a design procedure in an (n.n) - dimensional parameter space which is badly arranged and too complicated to be of much help in connection with complex practical problems.
The above announced compromise is found by writing the matrix K and the characteristic polynomial as

$$K = k.K_1 , \tag{7.3}$$

where k is a scalar variable ($k \in \mathbb{R}^1$) and K_1 is a constant matrix ($K_1 \in \mathbb{R}^{p,p}$) and

$$\Psi_c(s;k) = det(s.I - A + k.B.K_1) \tag{7.4.a}$$

Relation (7.3) implies that all elements of the matrix K are changed proportionally to each other. Using well known determinant identities, equation (7.4.a) may be reformulated as follows :

$$\Psi_c(s;k) = \det(s.I-A).\det(I+k.(s.I-A)^{-1}.B.K_1) \; , \qquad (7.4.b)$$

or

$$\Psi_c(s;k) = \det(s.I-A).\det(I+k.K_1.(s.I-A)^{-1}.B) \; , \qquad (7.4.c)$$

or

$$\Psi_c(s;k) = \det(s.I-A).\det(I+k.B.K_1.(s.I-A)^{-1}) \; . \qquad (7.4.d)$$

Using the abbreviations

$$\Psi_A(s) := \det\,(s.I-A) = \prod_{j=1}^{n_1} (s+s_{pj}) \qquad (7.5.a)$$

and

$$L := k.(sI-A)^{-1}.BK_1 = \frac{1}{\Psi_A(s)}.k.\mathrm{adj}^T(sI-A).BK_1 \; , \qquad (7.5.b)$$

yields

$$\Psi_c(s;k) = \Psi_A(s) \, . \, \det(I + L) \qquad (7.6.a)$$

or

$$\det\,(I + L) = \frac{\Psi_c(s)}{\Psi_A(s)} \; , \qquad (7.6.b)$$

where $\Psi_A(s)$ is the characteristic polynomial of the open-loop system (compare equation (7.5.a)), $L(s)$ is the return difference matrix, and $\det(I+L)$ is the dynamic control factor of the closed loop system, which constitutes a coarse measure of the effectivity of the control loop.

In single-loop control systems, L is a scalar variable, and the closed loop characteristic polynomial takes the form

$$\Psi_c(s;k) = \det(s.I-A+k.b.k_1^T)=\det((s.I-A).(I+k.(s.I-A)^{-1}.b.k_1^T), \qquad (7.7.a)$$

or using well-known determinant identities ,

$$\Psi_c(s;k) = \Psi_A(s;k).\det\,(1+k.k_1^T.(s.I-A).b) \; , \qquad (7.7.b)$$

or

$$\psi_c(s;k) = \psi_A(s;k) + k.k_1.adj^T(s.I-A).b = \prod_{i=1}^{n_1}(s+s_{pi}) + k.\prod_{j=1}^{m}(s+s_{nj}) \quad , \tag{7.7.c}$$

where $m \leq n_1$, and s_{pi} and s_{nj} are the poles and zeros of the open-loop transfer function, respectively.

Note :

Equations (7.7.a) to (7.7.c) are equivalent to equation (4.1). They can be investigated by classical root locus techniques as has been pointed out below. As is well known from the related construction rules, m of the closed-loop poles s_{pi} tend to the m zeros s_{nj} for increasing k, whereas s_{nj} of the (n-m) residual poles tend to infinity along asymptotes which constitute a Butterworth Polynomial.
Pole-zero compensations lead to root locus branches which are degenerated to an isolated point in the complex plane, which shows that the plant is either not completely controllable or not completely observable or both (compare Chapter VIII).
The characteristic equation of the linear multi-loop system

$$\psi_c(s;k) = \psi_A{}^{p-1}.\det(\psi_A.I + k.K_1.adj^T(sI-A).B) \tag{7.7.d}$$

is much more complex compared to equations(7.7.a) to (7.7.b). It cannot be discussed in complete analogy to those equations.
The special case of a quadratic regular open-loop transfer function matrix L(s) leads to the well known results of Theorem 7.1, which turn out to be a formal generalization of analogical statements in the scalar case (Compare [7.19]).

Theorem 7.1 :

Let the open-loop transfer function matrix L(s) of a linear multi-loop control system with p inputs and p outputs be regular.
Then the following statements hold :

a) The degree of the related McMillan normal form is equal

to p .

b) The control system has at most (n-p) zeros.

c) The zeros are invariant under state feedback.

The zeros are defined in the sense of McMillan in the above Theorem. On the basis of this definition, the related concepts of pole-zero-compensation and phase-minimum-system are easily carried over from linear single-loop control systems to regular multi-loop control systems, with all further control theoretical consequences.

But already in the regular case, there arise some difficulties in connection with these activities, [7.19]. Statement c) of Theorem 7.1 guarantees that the "McMillan zeros" of the system are invariant under state feedback, but it allows the zeros of the elements of the transfer function matrix of the closed loop system to be changed. From the point of view of practice, very offen, just these transfer functions are of primary interest. On the other hand, the transfer function matrix may not have a zero, though its elements may have various zeros.

This situation has been illustrated in Example 5.6 of Chapter V. The related transfer function matrices L(s), P(s) and $M_c(s)$, are regular. Neither the open loop system transfer function matrix nor the closed-loop system transfer function matrix have finite zeros. On the other hand, one of the scalar transfer functions of the open loop system has a zero $s=-2{,}5$ which is changed to $s=-2{,}5-0{,}5.k$ by state feedback. For negative values of k, the related transfer function may even become phase-non-minimal, though the overall system is phase-minimal (in the sense of McMillan).

For singular open-loop transfer function matrices, the analogy of the system zeros of single-loop control systems and of multi-loop-control systems with characteristic polynomials of the form (7.4.a) partially breaks down. As will be shown in the following discussions and examples, the zeros of singular multi-loop control systems with characteristic polynomials of the form (7.4.a) are no longer goals or end points of root locus branches for increasing k, contrary to single-loop control systems with characteristic polynomials of the form (7.7.a).

Starting from the closed-loop characteristic polynomial of equation (8.4.a), we have, using the abbreviation

$$R := - A + k.B.K_1 \quad : \tag{7.8}$$

$$\psi_c(s;k) = \det (sI + R) = \frac{1}{s^n}.\det (I + \frac{1}{s}.R) \tag{7.9.a}$$

and

$$\begin{aligned}\psi_c(s;k) = s^n.(1 &+ \frac{1}{s}.\ \text{trace}\ (R) \\ &+ \frac{1}{s^2}.\textstyle\sum \text{main minors of 2nd order of } R \\ &+ \vdots \\ &+ \frac{1}{s^i}.\textstyle\sum \text{main minors of ith order of } R \\ &+ \vdots \\ &+ \frac{1}{s^n}.\ \det R\ )\end{aligned} \tag{7.9.b}$$

or

$$\begin{aligned}\psi_c(s;k) = s^n &+ s^{n-1}\ .\ \text{trace}\ (R) \\ &+ s^{n-2}\ .\ \textstyle\sum \text{main minors of 2nd order of } R \\ &+ \vdots \\ &+ s^{n-i}.\ \textstyle\sum \text{main minors of ith order of } R \\ &+ \vdots \\ &+ \det R\ .\end{aligned} \tag{7.9.c}$$

Another representation of $\psi_c(s;k)$ is

$$\psi_c(s) = \det(s.I-A).\det(I + k.(s.I-A)^{-1}.B.K_1) \tag{7.10.a}$$

or using $\qquad L = k.M \quad , \quad L,M \in \mathbb{R}^{p,p}$

$$\psi_c(s;k) = \psi_A(s).\det (I + k.M) \tag{7.10.b}$$

and

$$\Psi_c(s;k) = \Psi_A(s) . \{ 1 + k . \text{trace}(M) \qquad (7.1o.c)$$
$$+ k^2 . \sum \text{main minors of 2 nd order of } M$$
$$\vdots$$
$$+ k^p \det M \} .$$

or

$$\Psi_c(s;k) = \Psi_A(s) + k . \Psi_A(s) . \text{trace}(M) \qquad (7.1o.d)$$
$$+ k^2 . \Psi_A(s) . \Sigma \text{main minors of 2nd order of } M$$
$$\vdots$$
$$+ k^p . \Psi_A(s) . \det M .$$

Using the abbreviations

$$M =: \frac{\bar{M}}{N} \quad \text{or} \quad \bar{M} := N.M \quad ,$$

where N is the smallest common multiple of the denominator polynomials of all rational elements of M , results in the expressions

$$\Psi_c(s;k) = \Psi_A . \{ 1 + k . \text{trace}(M) \qquad (7.1o.e)$$
$$+ k^2 . \sum_j H_{2j}$$
$$\vdots$$
$$+ k^{p-1} . \sum_j H_{p-1,j}$$
$$+ k^p . \det M \}$$

or

$$\Psi_c(s;k) = \Psi_A . \{ 1 + \frac{k}{N} . \text{trace}(\bar{M})$$
$$+ (\frac{k}{N})^2 \sum_j \bar{H}_{2j}$$
$$\vdots \qquad (7.1o.f)$$
$$+ (\frac{k}{N})^{p-1} . \sum_j \bar{H}_{p-1,j}$$
$$+ (\frac{k}{N})^p . \det \bar{M} \} ,$$

where $\bar{H}_{ij}$ and H_{ij} are main minors of order i of $\bar{M}$ and M, respectively, $H_p = \det M$ and $\bar{H}_p = \det \bar{M}$.

In case that the characteristic and the minimal polynomials

are identical, we have

$$\psi_A(s) = N \quad . \tag{7.11}$$

For simplicity, it is assumed in what follows, that Relation (7.11) holds. Then the McMillan normal form $M_c(s)$ of the matrix M may be written as

$$M_c = \frac{1}{N} \cdot \text{diag}\,(f_1, f_2, \ldots, f_p) = \frac{1}{N} \cdot S \quad , \tag{7.12}$$

where S is the Smith normal form of the matrix $\bar{M} = N.M$ and f_j are the invariant factors of $\bar{M}$:
Using the relations

$$D_i = \sum_{j=1}^{i} f_j \quad \text{for} \quad i = 1, \ldots, p \quad , \tag{7.13.a}$$

i.e.

$$f_1 = D_1 \ , \ f_2 = D_2 / D_1 = D_2 / f_1 \tag{7.13.b}$$

and

$$f_p = D_p / D_{p-1} = D_p \Big/ \prod_{j=1}^{p-1} f_j \quad , \tag{7.13.c}$$

where D_i are the greatest common divisors of all minors of order i of $\bar{M}$ (sometimes called determinant divisors of order i of $\bar{M}$), we have

$$D_i \mid D_{i+1} \quad \text{and} \quad f_i \mid f_{i+1} \quad .$$

On the basis of these relations, the following Definition 7.1 may be formulated :

<u>Definition 7.1 :</u>

Let $L = k.M$ be the open-loop transfer function matrix of a multi-loop control system under state feedback with feedback matrix K, where

$$K = k.K_1,\ k \in \mathbb{R}^1,\ K_1 \in \mathbb{R}^{p,n}\ (k \text{ variable, } K_1 \text{ constant}). \tag{7.14}$$

Let $H_r(s)$ be a non-vanishing main minor of maximum order r $(r \le p)$ of the matrix M.
Then we define :

(a) The roots of equation

$$\psi_A(s). \sum_j H_{rj}(s) = \psi_A(s)^{-r+1}. \sum_j \bar{H}_{rj}(s) \tag{7.15}$$

are called <u>sinks</u> of the closed-loop control system under state feedback $K = k.K_1$.

(b) The system has a <u>(partial) pole-sink-cancellation</u> if the expressions

$$\sum_j \bar{H}_{rj}(s) \quad \text{and} \quad \psi_A(s) \tag{7.16}$$

of equation (7.1o.f) have a common divisor of positive degree in s.
In this case , a geometric phenomenon, called "<u>loop fixation</u>", may occur in the related higher order root locus plot. (compare example 7.5).

(c) If there exists a common divisor of positive degree in s of all of the terms of equation (7.1o.f), there exist a <u>complete pole-sink-cancellation</u> of the closed loop system. In this case, at least one pole of the system transfer function matrix is not shifted under the above state feed-back.

(d) If at least one of the expressions $\bar{H}_{ij}(s) (1 \le i \le r-1)$ and $\bar{H}_r$ have a common root which is not a simultaneous root of $\psi_A(s)$, there exists a <u>secondary sink-cancellation</u> of the system.

(e) If at least one of the expressions $\sum_j \bar{H}_{ij}(s)$, $(1 \le i \le r-1)$, and $\psi_A(s)$ have a common root which is not a simultaneous root of $H_r(s)$,there exists a secondary <u>pole-sink-cancellation</u> of the system.

<u>Note :</u>

(i) On the basis of Definition 7.1, various further control theoretical concepts may be introduced, as for instance a sink-phase-minimal multi-loop control system , ... This will be omitted here.

(ii) As will be shown in a later paper , the concept of a sink is not only useful in connection with linear control systems but also in the analysis and design of <u>nonlinear</u> control systems.

Using the concepts of Definition 7.1,the following statements can be proved :

<u>Theorem 7.2 :</u>

Given a linear multi-loop control system under state feedback of the form $K = k.K_1$ with characteristic polynomial (7.1o.f) Then the following statements hold :

(a) The divisibility condition

$$D_i \Big| \sum_j \bar{H}_{ij} \quad \text{for} \quad i = 1,2, \ldots ,p \qquad (7.17)$$

holds, where D_i and $\bar{H}_{ij}$ are defined above.

(b) If the multi-loop control system has r sinks, then r branches of the related higher order root locus plot tend towards these sinks for $k \to \infty$ (motivation of the name "<u>sink</u>").

(c) Let the open loop transfer function matrices L(s) and M(s) be <u>regular.</u> Then the sinks of the feedback system are identical to the zeros of the system in the sense of McMillan. If the matrices L(s) and M(s) are singular, then the system zeros usually differ from the system sinks (compare example 7.3).

(d) If the terms $\psi_A(s)$ and $\sum_j \bar{H}_{rj}(s)$ have a common divisor q(s) of positive degree in s, the system has a (partial) pole-sink-cancellation (compare example 7.5).

If q(s) divides at the same time all other terms of (7.1o.f), there exist a complete pole-sink-canellation (compare examples 7.1, 7.3 and 7.4).

If q(s) is not a simultaneous divisor of all other terms of (7.1o.f), there may exist a "loop-fixation" of the higher order root locus plot.

Proof :

(a) According to (7.13.a) to (7.13.c), D_i is the greatest common divisor of all minors of order i of $\bar{M}$. Then D_i also divides all main minors $\bar{H}_{ij}$ of order i of $\bar{M}$ and the expression $\sum_j \bar{H}_{ij}$.

(b) Let $\bar{H}_{rj}(s)$ be a non vanishing minor of $\bar{M}$ of maximum order r, where $r \leq p$.
For $r < p$ we have, using (7.1o.f) and the relations $\bar{H}_{qj}(s) = o$ for $q > r$, and $\det \bar{M}(s) = H_p(s) = o$:

$$\psi_c(s;k) = \psi_A \cdot \{1 + \frac{k}{N} \cdot \mathrm{trace}(\bar{M}) + (\frac{k}{N})^2 \cdot \sum_j \bar{H}_{2j} + \ldots + (\frac{k}{N})^r \cdot \sum_j \bar{H}_{rj}\} \qquad (7.18.a)$$

and

$$\lim_{k \to \infty} k^{-r} \cdot \psi_c(s;k) = \psi_A^{-r+1}(s) \cdot \sum_j \bar{H}_{rj} \quad . \qquad (7.18.b)$$

The roots of equation (7.18.b) are called sinks in Definition 7.1 .

(c) From r = p we have :

$$\lim_{k \to \infty} k^{-p} \cdot \psi_c(s;k) = \det M = \psi_A(s)^{-p+1} \cdot \det \bar{M} \; . \qquad (7.19.a)$$

The roots of (7.19.a) are identical to the zeros of M(s) in the sense of McMillan. They are roots of the relation

$$D_p(s) = \prod_{i=1}^{p} f_i(s) \quad . \qquad (7.19.b)$$

This follows from the well known fact, that the characteristic polynomials of unimodular equivalent systems have identical roots.

(d) Statements (d) of Theorem 7.2 follow directly from equations (7.1o.e) and (7.1o.f) in connection with Definition 7.1 .

Note :

Relation (7.17) shows that the knowledge of the structural invariants (i.e. D_j or f_i) of M(s) only provide enough information to determine the sinks of a multi-loop control system in case of a regular matrix M(s) or in the special case, that all greatest common divisors of the main minors of M(s) and of $\bar{M}(s)$ are identical, respectively.
As is easily seen by comparing the results of Theorem 7.2 and of Equations (4.43) and (4.44) of Chapter IV, the various concepts of pole-sink-cancellations are closely related to the return points of the corresponding higher root locus plots.

Linear completely controllable and completely observable control plants have transfer functions or transfer function matrices with no pole-zero-cancellations. As is well known, single-loop control systems (with k as feedback factor) have degenerated root locus branches, iff the related open-loop transfer function has pole-zero cancellations. In this case, the existence of degenerated root locus branches is equivalent to a loss of complete controllability and/or observability of the open loop system.

In case of linear multi-loop control systems with state feedback of the form (7.3) (e.g. $K = k.K_1$), the existence of a degenerated branch of the related higher order root locus plot is equivalent to a complete pole-sink cancellation of the system.

Equation (7.3) with $K = k.K_1$ provides a rather specific feedback control law. There may occur pole-sink-cancellations even for control systems which are completely controllable and observable.

This gives rise to the following definition :

Definition 7.2:

A linear control system is completely K_1- controllable under state feedback (7.3) ($K = k.K_1$), iff there exist no pole-sink cancellations of the closed loop system.

The following corollary 7.3 is a trivial consequence of the above discussion.

Corollary 7.3:

A linear system is not completely K_1- controllable under state feedback (7.3) iff the related higher order root locus plots have degenerated branches.

2. Examples

Example 7.1 : (Linear multiloop system ; singular transfer function matrix).

Given the multi-loop system of example 5.6, with singular feedback matrix

$$K = k.\, K_1 \;,\; K_1 = \begin{pmatrix} 1 & 0 & 0 \\ 0 & 1 & 1 \\ 0 & 1 & 1 \end{pmatrix}, \text{ degree } k_1 = 2 \;.$$

Then the open-loop system has the transfer function matrix

$$L(s) = P(s).K =$$

$$\frac{k}{(s+1).(s+2).(s+3)} \cdot \begin{pmatrix} (s+2).(s+3) & 3.(s+8/3) & 3.(s+8/3) \\ 0 & (s+1).(s+4) & (s+1).(s+4) \\ 0 & (s+1).(s+2) & (s+1).(s+2) \end{pmatrix}$$

and the McMillan form

$$M_{CL}(s) = k. \begin{pmatrix} \frac{1}{(s+1).(s+2).s+3} & 0 & 0 \\ 0 & 1 & 0 \\ 0 & 0 & 0 \end{pmatrix}$$

related to it. As a result, the transfer function matrix of the open loop system is <u>singular.</u>

The closed loop system has the transfer function matrix

$$G_c(s;k) =$$

$$= \begin{pmatrix} \frac{1}{s+k+1} & \frac{s+3-k}{\psi_c} & \frac{2.s+k+5}{\psi_c} \\ 0 & \frac{s+3+k}{(s^2+5s+6)+k.(2.s+6)} & \frac{1-k}{(s^2+5.s+6)+k.(2.s+6)} \\ 0 & \frac{-k}{(s^2+5s+6)+k.(2s+6)} & \frac{(s+2+k)}{(s^2+5s+6)+k.(2.s+6)} \end{pmatrix} = \frac{1}{\psi_c} \cdot$$

$$\cdot \begin{pmatrix} (s+3).(s+2+2k), & (s+3-k) & (2s+k+5) \\ 0 & (s^2+4s+3)+k.(2s+4)+k^2 & (s+1-k.s-k^2) \\ 0 & -k.(s+1)-k^2 & (s^2+3s+2)+k.(2s+3)+k^2 \end{pmatrix}$$

and the characteristic polynomial

$$\psi_c(s;k) = (s+3).\{(s+1).(s+2)+k.3.(s+4/3)+k^2.2\}$$

or

$$\psi_c(s;k) = (s^3+6.s^2+11.s+6)+k.(3.s^2+13.s+12)+k^2.(2.s+6)$$

Construction of the higher order root locus plot.

Rule 1 : Polynomial coefficients of $Q_j(s)$ and $R_i(k)$.

	s^0	s^1	s^2	s^3	
Q_2	6	2			k^2
Q_1	12	13	3		k^1
Q_0	6	11	6	1	k^0
	R_0	R_1	R_2	R_3	

Table 7.a.1: Polynomial coefficients of $Q_j(s)$ and $R_i(k)$.

Rule 2 : Roots of $Q_j(s)$ and $R_i(k)$.

Q_j	i	$s-s_1^i$	$s-s_2^i$	$s-s_3^i$	
Q_2	2	(s+3)			k^2
Q_1	1	(s+3)	(s+4/3)		k^1
Q_0	o	(s+1)	(s+2)	(s+3)	k^0
		R_0	R_1	R_2	

Table 7.b.1 : Roots of $Q_j(s)$.

	s^0	s^1	s^2	
Q_1	k+1	k+1		$k-k_j^1$
Q_0	k+1	k+5,5	k+2	$k-k_j^o$
j	o	1	2	
R_j	R_o	R_1	R_2	

Table 7.c.1: Roots of $R_i(k)$

Rules 3 and 4 : Exponent Diagram

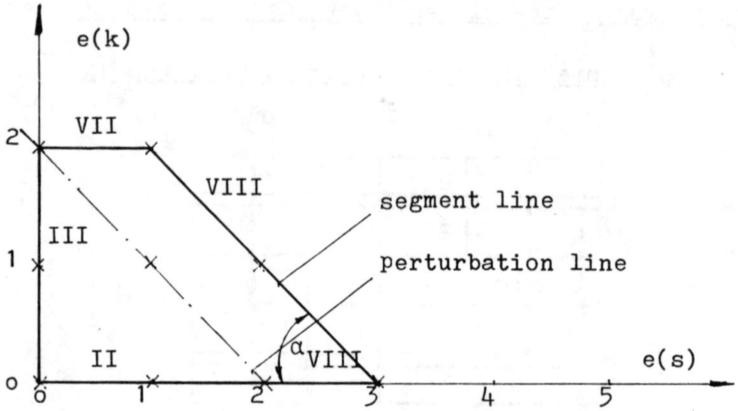

Figure 7.1.a : Exponent Diagram of P(s;k)

Rule 5 : Constants

$l_o = 0$, $n_o = n = 3$, $q = 2$

$e_{II} = 3$, $e_{VIII} = 2$, $e_{VII} = 1$, $\bar{e}_{III} = 2$

Rule 7 : Slopes

$\beta_{VIII} = 2/2 = 1$

Note :

There exist three root locus branches one of which is degenerated to the point s=-3, and two of which tend to infinity for increasing k, k> o . There exists a complete pole-sink-cancellation for s=-3.
The system is not completely K_1-controllable under k.
A pole-zero cancellation does not exist as the transfer function matrix does not have a zero.

Rule 11 : Angles of departure from $s_j \neq 0$ (segment II, k = o)

Replacing the variable s of Q_0, Q_1 and Q_2 by the variables $s' + s_j^o$ yields for

$s_1^o = -1$:

$$Q_0'(s') = 0 + 2s' + \dots$$

$$Q_1'(s') = 2 + \dots$$

$$\gamma = -1 \quad , \quad \beta = 1$$

$$\varphi(-1,0) = \pi \qquad \text{for} \quad k > 0 ,$$

and for

$s_2^o = -2$:

$$Q_0'(s') = 0 - 1.s' + \dots$$

$$Q_1'(s') = -2 + \dots$$

$$\gamma = -0{,}5 \quad , \quad \beta = 1$$

$$\varphi(-2,0) = \pi \qquad \text{for} \quad k > 0 \quad .$$

Rule 14 : Asymptote angles (segment VIII, $k = \infty$).

Using $\beta_{VIII} = 1$, the supporting polynomial

$$\gamma^2 + 3 \cdot \gamma + 2 = 0 ,$$

with roots $\gamma_1 = -1$ and $\gamma_2 = -2$,

and equation (4.27.b) yields

$$\varphi_1(\infty,\infty) = \pi$$
$$\varphi_2(\infty,\infty) = \pi \qquad \text{(for } k > o\text{)}.$$

The root locus plot is drawn in Figures 7.1.b and 7.1.c .

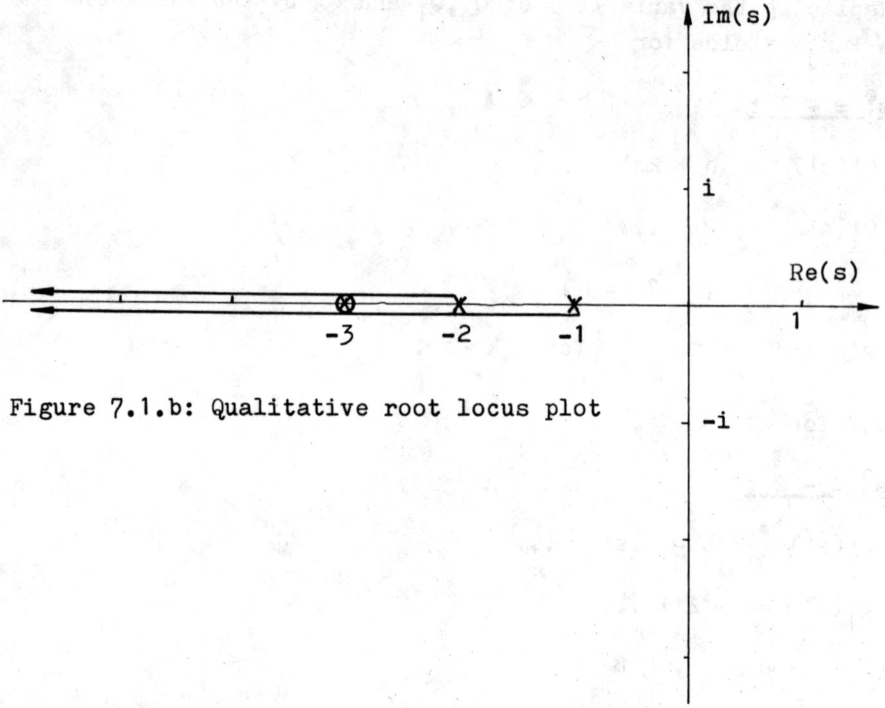

Figure 7.1.b: Qualitative root locus plot

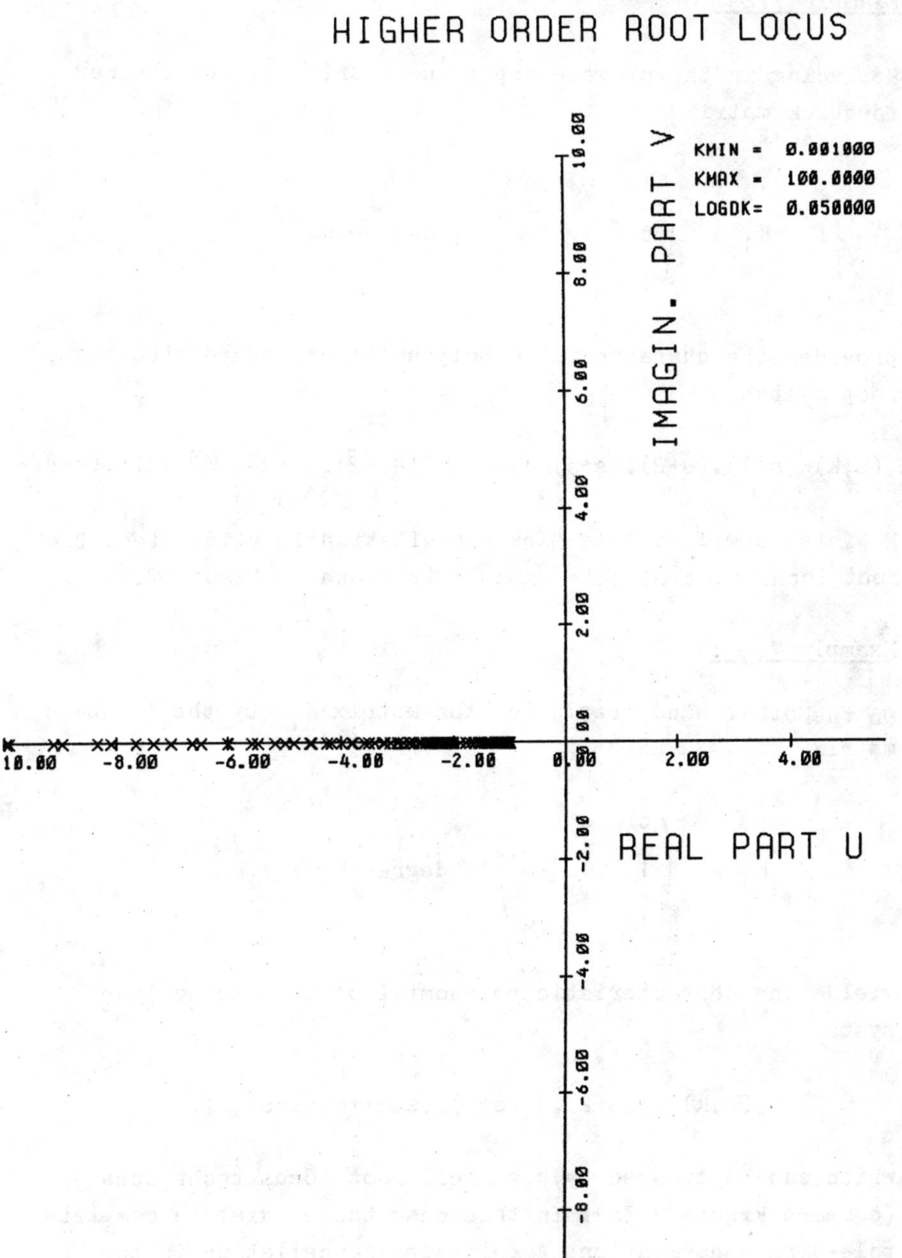

Figure 7.1.c: Computer root locus plot .

Example 7.2 :

Replacing in the above example the matrix K_1 by the new feedback matrix

$$K_1 = \begin{pmatrix} 1 & 0 & 0 \\ 0 & 1 & 2 \\ 0 & 1 & 2 \end{pmatrix}, \text{ degree } K_1 = 2 ,$$

provides the characteristic polynomial of the related closed loop system

$$\psi_c(s;k)=(s+1).(s+2).(s+3)+k.4.(s+2+\tfrac{i}{2}.\sqrt{2}).(s+2-\tfrac{i}{2}.\sqrt{2})+k^2.3.(s+8/3)$$

Now, the complete pole-sink-cancellation is eliminated. The root locus plot of this example is shown in Figure 7.2.a .

Example 7.3 :

On the other hand, replacing the matrix K_1 by the feedback matrix

$$K_1 = \begin{pmatrix} 0 & 0 & 0 \\ 1 & 1 & 0 \\ 1 & 1 & 0 \end{pmatrix}, \text{ degree } K_1 = 1$$

yields the characteristic polynomial of the closed loop system

$$\psi_c(s;k) = (s+2).[(s+1).(s+3)+k.(s+6)] ,$$

which can be treated by classical root locus techniques (compare Figure 7.3a). In this case there exists a complete pole-sink-cancellation. A pole-zero-cancellation in the sense of McMillan does not exist because the McMillan normal form of the open loop system transfer function L(s) does not have a zero. The system is not completely K_1-controllable.

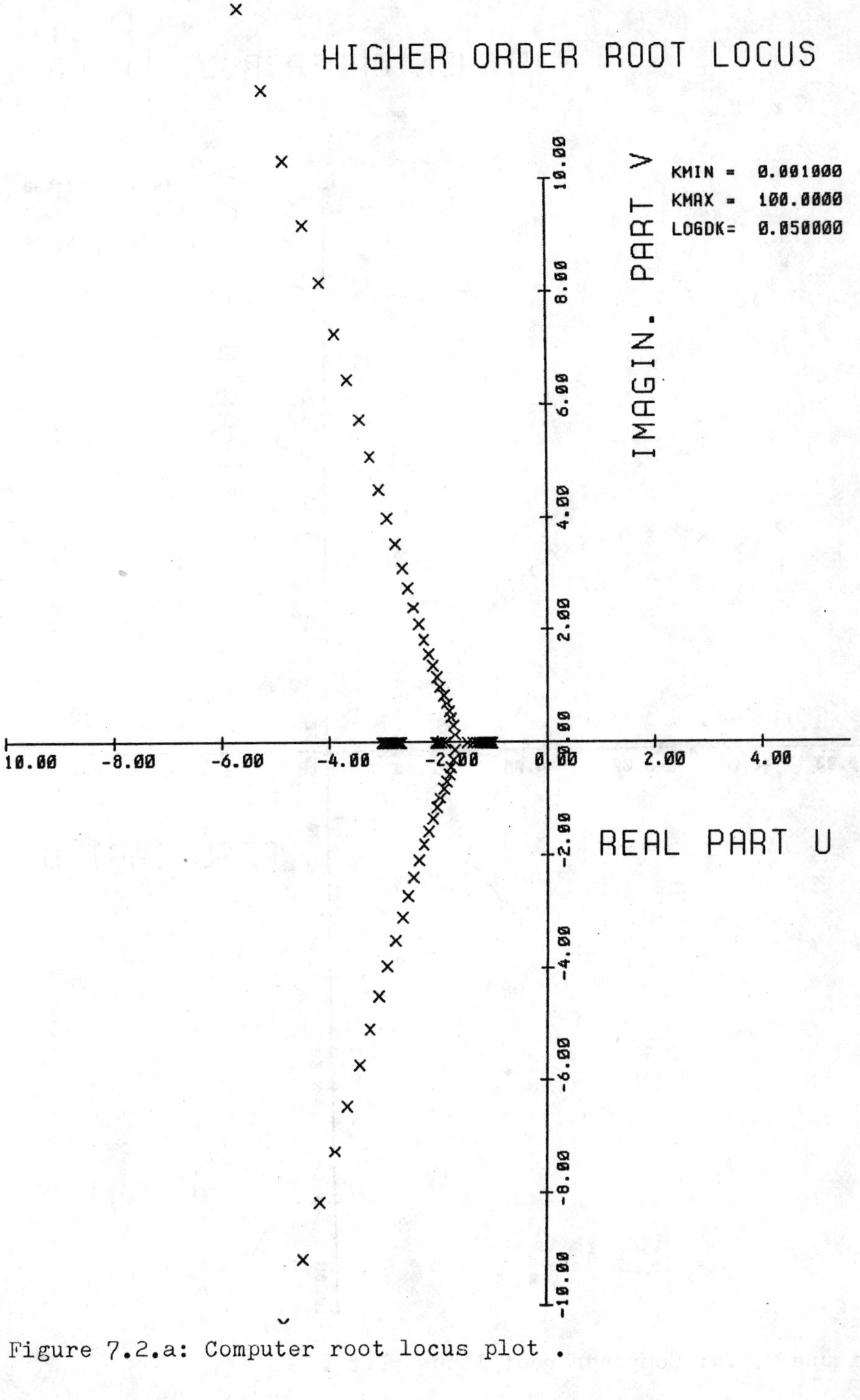

Figure 7.2.a: Computer root locus plot .

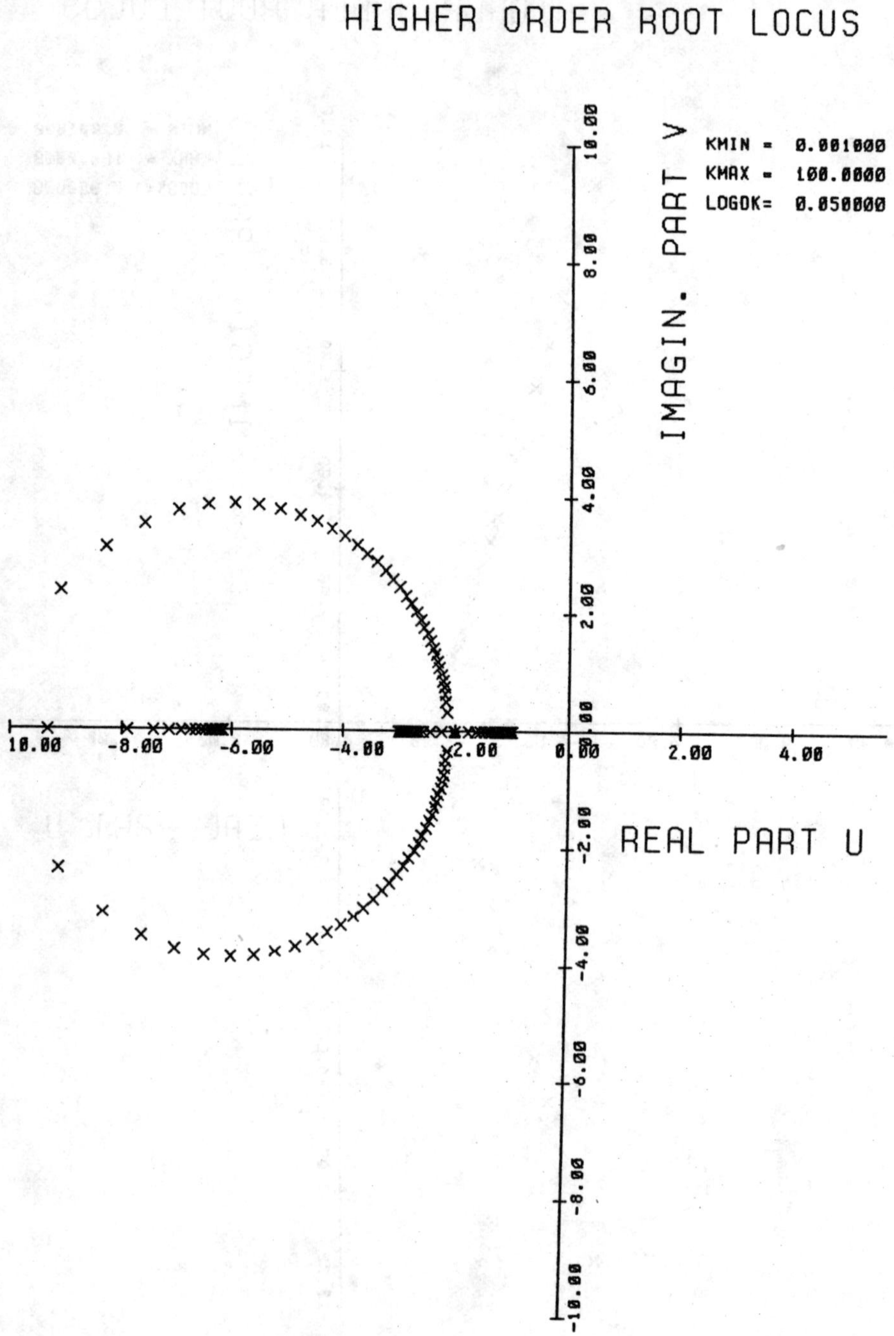

Figure 7.3.a: Computer root locus plot .

Example 7.4 :

Choosing the matrices B and K_1 of example 5.6 as

$$B = \begin{pmatrix} 1 & 0 \\ 0 & 1 \\ 0 & 0 \end{pmatrix} , \quad K_1 = \begin{pmatrix} 1 & 0 & 0 \\ 0 & 1 & 0 \end{pmatrix}$$

yields the open loop transfer function matrix

$$L = \frac{k}{(s+1).(s+2).(s+3)} . \begin{pmatrix} (s+2).(s+3) & (s+3) & 0 \\ 0 & (s+1).(s+3) & 0 \\ 0 & 0 & 0 \end{pmatrix}$$

$$= k. \begin{pmatrix} \frac{1}{(s+1)} & \frac{1}{(s+1).(s+3)} & 0 \\ 0 & \frac{1}{(s+2)} & 0 \\ 0 & 0 & 0 \end{pmatrix} ,$$

the corresponding McMillan normal form

$$M_{CL} = k. \begin{pmatrix} \frac{1}{(s+1).(s+2)} & 0 & 0 \\ 0 & 1 & 0 \\ 0 & 0 & 0 \end{pmatrix} ,$$

and the related closed loop characteristic polynomial

$$\psi_c(s;k) = \det (sI - A + B K)$$

$$= (s+3). \left\{ (s+1).(s+2)+2.k.(s+1,5)+k^2 \right\} .$$

There exists a complete pole sink compensation at $s = -3$.
The system is not completely K_1-controllable.
The relation

$$rg\ (B,\ A.B,\ A^2.B) = rg\ (B.K_1,\ A.B.K_1,\ A^2.B.K_1) = 2 < 3.$$

shows that the system is not completely controllable.
The root locus plot is drawn in Figures 7.4.a and 7.4.b .

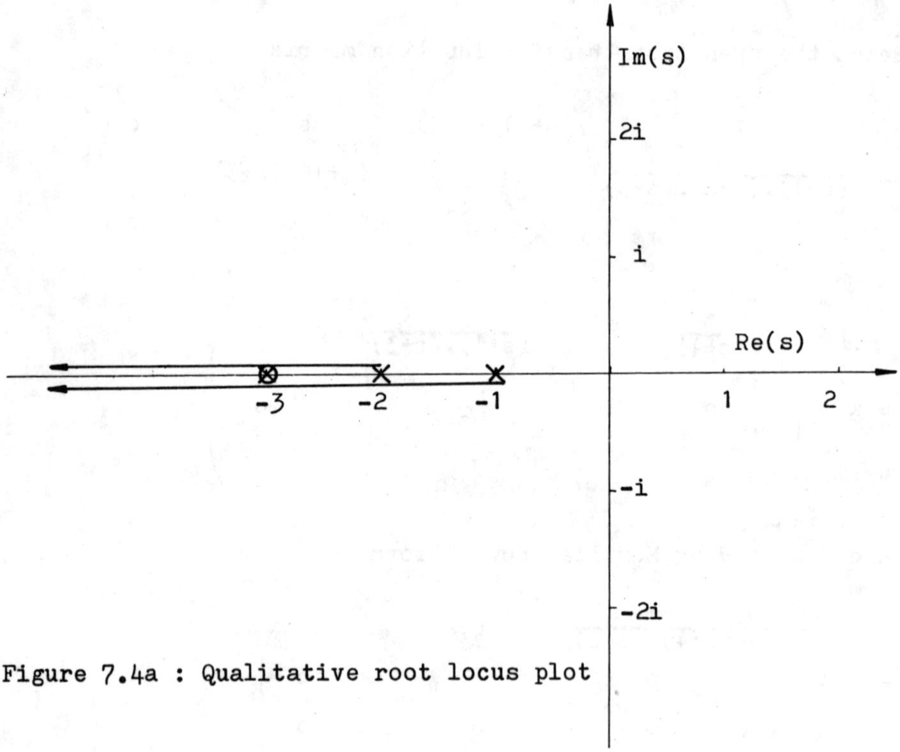

Figure 7.4a : Qualitative root locus plot

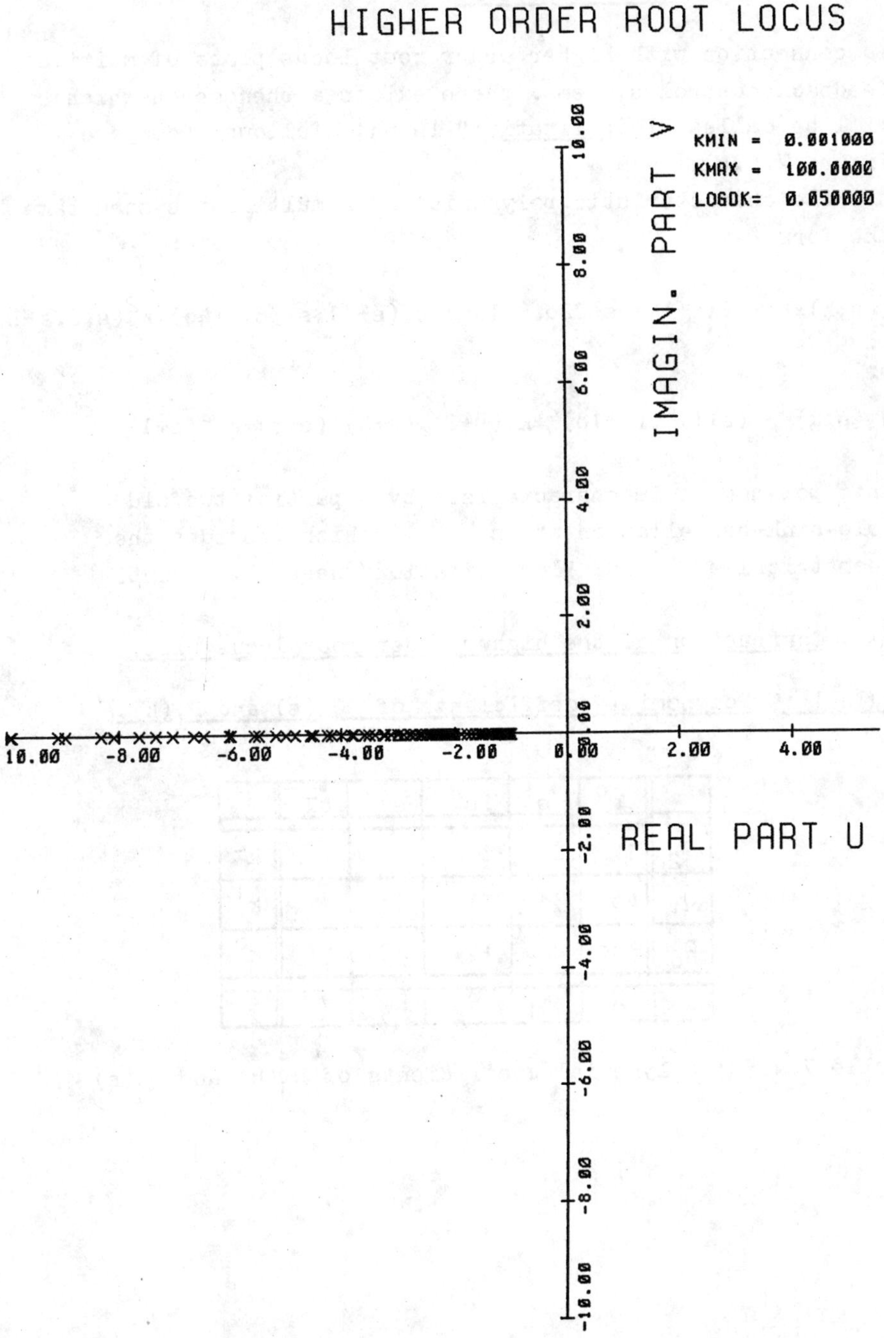

Figure 7.4.b: Computer root locus plot .

Example 7.5 : <u>(Loop fixation)</u>

In connection with higher order root locus plots of multiloop feedback control systems, there exists a phenomenon which will be called "loop fixation" in what follows (compare Figure 7.5.b).
Let the characteristic polynomial of a multiloop system take the form

$$\psi_c(s;k)=(s^4+22.s^3+141.s^2+220.s+100)+k.(s^3+11.s^2+38.s+40)+k^2.(s^2+2.s+1)$$

or

$$\psi_c(s;k) = (s+1)^2.(s+10)^2+k.(s+2).(s+4).(s+5)+k.^2(s+1)^2 .$$

This polynomial is characterized by a partial twofold pole-sink-cancellation at $s = -1$ which provides the geometrical effect of "loop fixation" near this point.

Contruction of the higher order root locus plot.

Rule 1 : Polynomial coefficients of $Q_j(s)$ and $R_i(k)$.

	s^0	s^1	s^2	s^3	s^4	
Q_2	1	2	1			k^2
Q_1	40	38	11	1		k^1
Q_0	100	220	141	22	1	k^0
	R_0	R_1	R_2	R_3	R_4	

Table 7.a.5 : Polynomial coefficients of $R_j(k)$ and $Q_j(s)$

Rule 2 : Roots of polynomials $Q_j(s)$ and $R_i(k)$

Q_j	i	$s-s_1^i$	$s-s_2^i$	$s-s_3^i$	$s-s_4^i$	
Q_2	2	(s+1)	(s+1)			k^2
Q_1	1	(s+2)	(s+4)	(s+5)		k^1
Q_0	o	(s+1)	(s+1)	(s+1o)	(s+1o)	k^0
		R_0	R_1	R_2	R_3	

Table 7.b.5 : Roots of $Q_j(s)$

	s^0	s^1	s^2	s^3	
Q_1	(k+37,3)	k+9,5 +i.4,5			$k-k_j^1$
Q_0	(k+2,7)	k+9,5 -i.4,5	(k+12,8)	(k+22)	$k-k_j^o$
j	o	1	2	3	
R_j	R_0	R_1	R_2	R_3	

Table 7.c.5 : Roots of $R_i(s)$

Rules 3 and 4 : Exponent Diagram

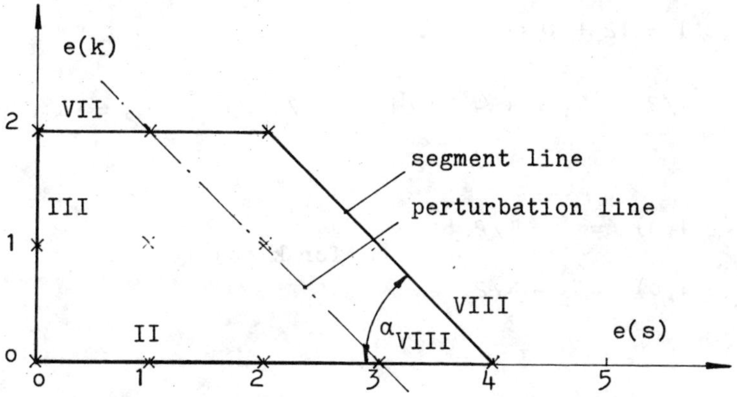

Figure 7.5a: Exponent diagram

Rule 5 : Constants

$l_0 = 0$, $n = n_0 = 4$, $q = 2$

$e_{II} = 4$, $e_{VIII} = 2$, $e_{VII} = 2$, $\bar{e}_{III} = 2$.

Rule 7 : Slopes

$$\beta_{VIII} = 2/2 = 1$$

Note :

There exist 4 root locus branches, two of which tend to the sink $s = -1$, and the others tend to infinity for increasing k ($k > o$). There exists a double partial pole-sink-cancella - tion at $s = -1$. The system is completely K_1-controllable.

Rule 11 : Angles of departure from $s_j^o \neq o$, $j = 1,2,3$ (segment II, k = o).

Replacing the variable s of $Q_0(s)$ and $Q_1(s)$ by the variables $s'-1$ and $s'-1o$ yields for

$s_1^o = -1$:

$$Q_0'(s') = o + o.s' + 81.s'^2 + \dots ,$$

$$Q_1'(s') = 12 + 19s' + \dots ,$$

$$\beta_I = 1/2 , \quad \gamma_1 = +\sqrt{-12/81} = i.2/9.\sqrt{3} , \quad \gamma_2 = -i.2/9.\sqrt{3}$$

and

$$\varphi_1(-1,o) = \pi/2$$

$$\varphi_2(-1,o) = -\pi/2 , \qquad \text{(for } k > o\text{)}$$

for

$\underline{s_1^o = -10:}$

$$Q_o'(s') = o + o.s' + 81.s'^2 + \ldots \quad ,$$

$$Q_1'(s') = -240 + 118.s' + \ldots \quad ,$$

$$\beta_I = 1/2 \quad , \quad \gamma_1 = +(240/81)^{1/2} = 1{,}72 \; , \; \gamma_2 = -1{,}72$$

and

$$\left.\begin{aligned} \varphi_1(-10,o) &= o \\ \varphi_2(-10,o) &= \pi \end{aligned}\right\} \quad \text{(for } k > o\text{)} \quad .$$

Rule 13 : Angles of arrival at $s_1^2 = -1$ (segment VII, $k = \infty$).

Replacing the variable s of $Q_1(s)$ and $Q_2(s)$ by the variable $s'-1$ yields

$$Q_1'(s) = 12 + 19s' + \ldots \quad ,$$

$$Q_2'(s) = o + o.s' + 1.s'^2 + \ldots \quad ,$$

$$\beta_V = 1/2 \quad , \quad \gamma_1 = i.\sqrt{12} \; , \quad \gamma_2 = -i.\sqrt{12} \quad ,$$

$$\left.\begin{aligned} \varphi_1(-10,\infty) &= \pi/2 + \pi = 3/2.\pi = -\pi/2 \\ \varphi_2(-10,\infty) &= -\pi/2 + \pi = +\pi/2 \end{aligned}\right\} \quad \text{(for } k > o\text{)} \quad .$$

Rule 14 : Asymptote angles (segment VIII, $k = \infty$).

Using $\beta_{VIII} = 1$, $\gamma^2 + \gamma + 1 = o$ or

$$\gamma_1 = -0{,}5 + i.0{,}5.\sqrt{3} \quad , \quad \gamma_2 = -0{,}5 - i.0{,}5.\sqrt{3}$$

yields in connection with (4.27b) :

$$\left.\begin{aligned} \varphi_1(\infty,\infty) &= 12o^o \\ \varphi_2(\infty,\infty) &= -12o^o \end{aligned}\right\} \quad \text{(for } k > o\text{)}.$$

Rule 15 : Asymptote points

Relation (4.37a) in connection with γ_1 and γ_2 from Rule 14 and Figure 7.5.a yields the expression

$$s_{o_j} = \frac{22.\gamma_j^3 + 11.\gamma_j^2 + 2.\gamma_j}{4.\gamma_j^3 + 3.\gamma_j^2 + 2.\gamma_j} \quad , \quad j = 1,2 \quad ,$$

or

$$s_{o_1} = -9{,}94 + i.\,0{,}6$$

and

$$s_{o_2} = -9{,}94 - i.0{,}6 \quad .$$

The above rules show that there exist two real return points $s_{R1} < -10$ and $s_{R2} > -10$.

The root locus plots are shown in Figures 7.5.b and 7.5.c . There exist two loopes of complex conjugate root locus branches starting at $s = -1$ for $k = 0$ and ending at $s = -1$ for $k = \infty$.

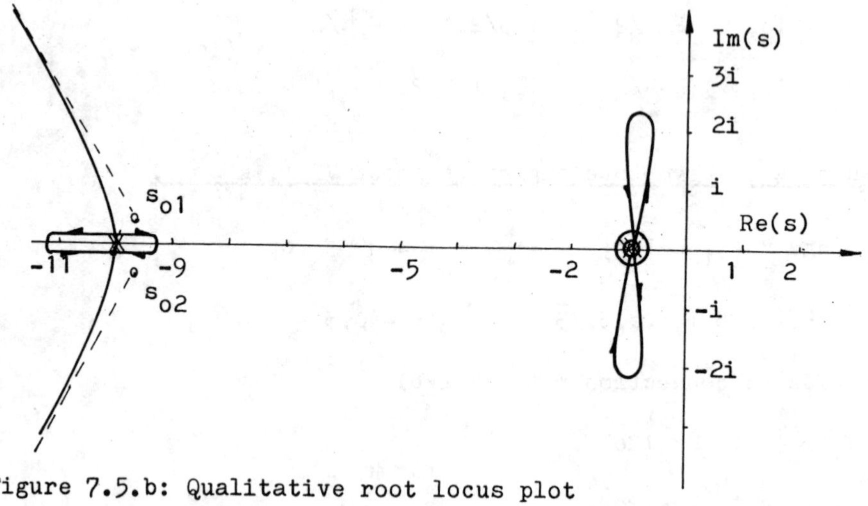

Figure 7.5.b: Qualitative root locus plot

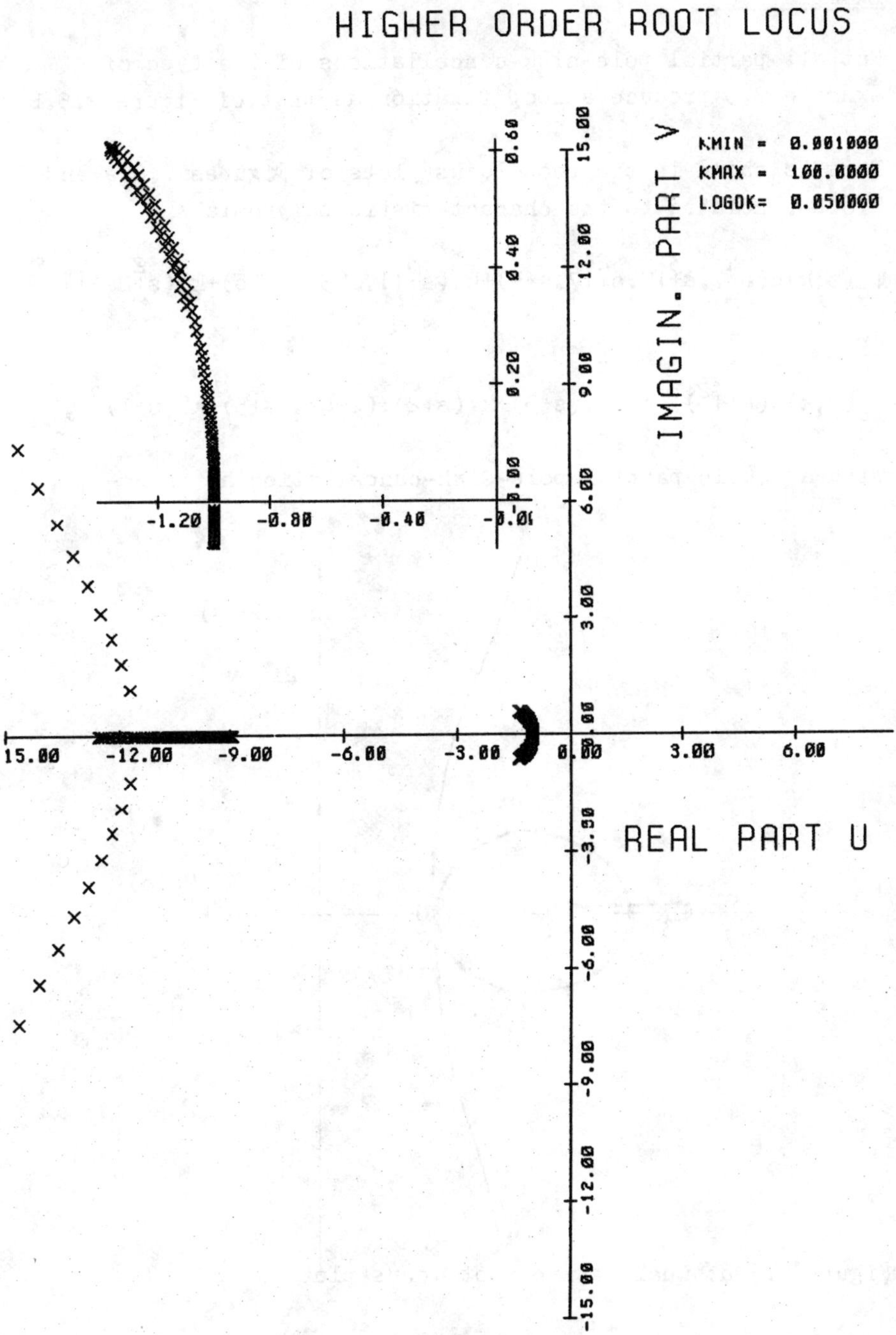

Figure 7.5.c: Computer root locus plot.

Example 7.6 :

Not all partial pole-sink-cancellations of the type of example 7.5 produce a loop fixation as that of Figure 7.5.b .

This is shown in the root locus plots of Figures 7.6.a and 7.6.b , related to the characteristic polynomial

$$\psi_c(s;k)=(s^4+7.s^3+17.s^2+17.s+6)+k.(s^3+11.s^2+38.s+40)+k^2.(s^2+2s+1)$$

or

$$\psi_c(s;k)=(s+1^2).(s+2).(s+3)+k.(s+2).(s+4).(s+5)+k^2.(s+1)^2 ,$$

with a double partial pole-sink-cancellation at $s = -1$.

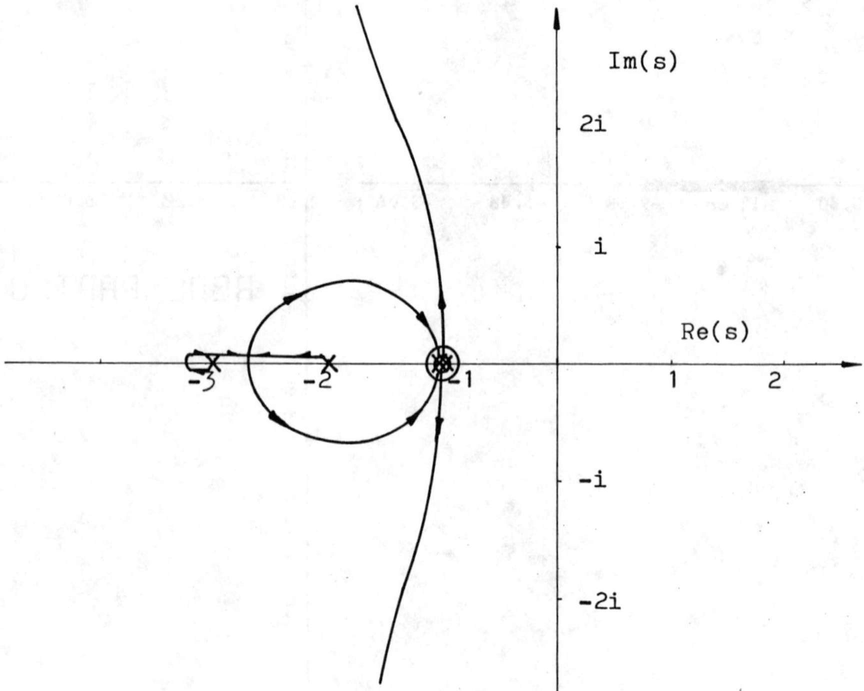

Figure 7.6.a: Qualitative root locus plot

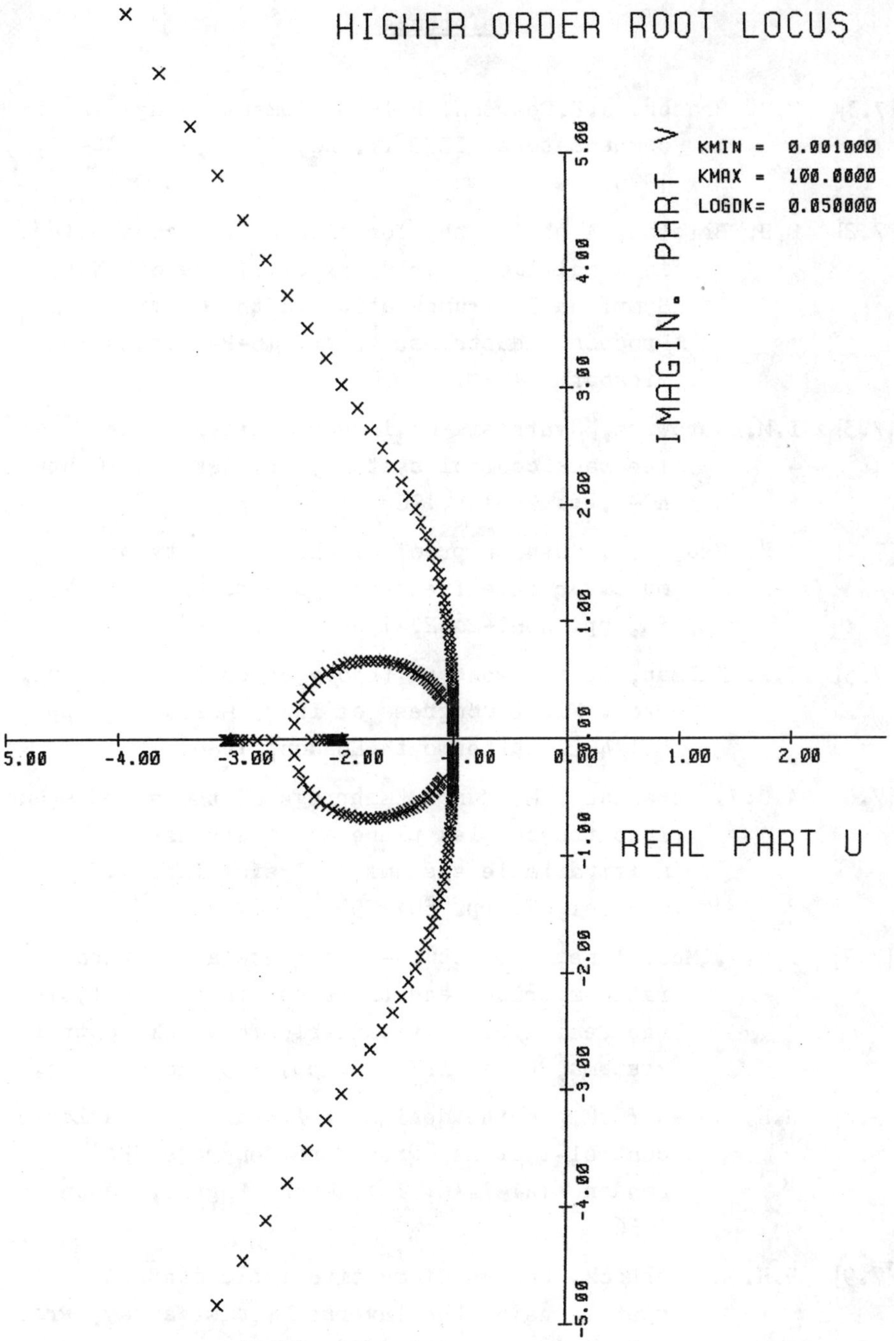

Figure 7.6.b: Computer root locus plot.

References

[7.1] F.M. Brasch, J.B.Pearson, Pole placement using dynamic compensators, IEEE Tr. AC, AC-15, pp. 34-43, 197o.

[7.2] E.H. Bristol, A philosophy for single loop controllers in a multiloop world, Proc. of the 8th Nat. Symp. on Instrumentation in the Chemical and Process Industries, 4, pp. 18-29, St.Louis, Missouri ,1967.

[7.3] I.M. Horowitz, Synthesis of linear multivariable feedback control systems, Tr. IRE Aut. Control AC-5,pp. 94-1o5,196o.

[7.4] C.H. Hsu, C.T. Chen, A proof of the stability of multivariable feedback systems, Proc. IEEE, 56, pp. 2o61-2o62, 1968.

[7.5] R.E. Kalman, On the general theory of control systems, Proc. First Congress of IFAC, Moscow, 1, pp. 481-492, Butterworth London, 196o.

[7.6] A.G.J. McFarlane, M. Munro, Mappings of the state space into the complex plane and their use in multivariable systems analysis, Int. J. Control, 7, pp. 5o1-555, 1968.

[7.7] A.G.J. McFarlane, The return-difference and return-ratio matrices and their use in the analysis and design of multivariable feedback control systems, Proc. IEE 117, pp. 2o37-2o49, 197o.

[7.8] H.H. Rosenbrock, On the design of linear multivariable control systems, Proc. 3rd Congress IFAC London, 1A1-1A16, Int. Mech. Engrs.,London 1966.

[7.9] H.H. Rosenbrock, Design of multivariable control systems using the inverse Nyquist array, Proc. IEE, 116, pp. 1929-1936, 1969.

[7.1o] A.G.J. McFarlane, A Survey of some Recent Results in Linear Multivariable Feedback Theory, Automatica, Vol. 8, pp. 455-492, 1972.

[7.11] E.J. Davison, W.M.Wonham, On pole assignment in multivariable linear systems, IEEE Tr.,AC-12, pp. 66o-665, 1967.

[7.12] A.G.J. McFarlane, N. Karcanias, Poles and zeros of linear systems, a survey of the algebraic, geometric and complex variable theory, Int. J. Control, Vol 24, No 1, pp.35-74, 1976.

[7.13] F. Fallside, Control System Design by Pole Zero Assignment, Academic Press, London, 1977.

[7.14] B. A. Francis, W.M. Wonham, The Role of transmission zeros in linear multivariable regulators, Int. J. Contr. , Vol. 22, No. 5, pp. 678-681, 1975.

[7.15] H.H. Rosenbrock, The zeros of a system, Int.J.Control Vol. 18, No.2, pp. 297-299, 1973 and Vol.2o No.3, pp.525-537,1974.

7.16 G. Roppenecker, Nullstellen von Mehrgrössensystemen, Definitionen, Interpretationen und Berechnung, Diplomarbeit, Institut für Regelungstechnik, Universität Karlsruhe,1978.

[7.17] H. Hahn, Wurzelortskurven höherer Ordnung und Senken in linearen Mehrgrössenregelsystemen,Preprint Universität Tübingen, Fachbereich Physik, 1975/76.

[7.18] H. Hahn, Zur Theorie und Technik singulärer Regelkreise, Habilitationsschrift Universität Tübingen,Fachbereich Physik,1977/78.

[7.19] H. Schwarz, Optimale Regelung linearer Systeme, BI-Verlag, 1976.

VIII. Synthesis of Optimal Control Systems.

Given a linear time invariant system of the form

$$\dot{x} = A.x + B.u \quad , \quad y = \bar{C}.\, x$$
$$x \in \mathbb{R}^n \; ; \; u,y \in \mathbb{R}^p \quad , \quad \text{with} \tag{8.1}$$

x state vector, u input vector and y output vector.

Find a control law $u^o(t,x(t))$ which is capable of steering the system from any given state $x(t_o)$ at initial time t_o to any desired state $x(t_e)$ at a final time t_e and which minimizes the quadratic optimization criterion

$$J = \int_0^{t_e} (y^T.Q.y + u^T.R.u).dt \quad , \tag{8.2a}$$

where

$$t_o = o \; , \; Q := Q^T = C_1^T.C_1 \geq o \quad \text{and} \quad R = R^T > o \; . \tag{8.2b}$$

Let the system (A,B) be completely controllable, i.e .

$$\text{rg}\; Q_C := \text{rg}\; (B,\; A.B,\; A^2.B,\; \ldots\; ,\; A^{n-1}.B) = n \tag{8.3a}$$

and let the system $(A,\bar{C})$ be completely observable , i.e .

$$\text{rg}\; Q_o^T := \text{rg}\; (\bar{C}^T, A^T.\bar{C}^T\; ,\; \ldots\; , A^{T^{n-1}}.\bar{C}^T) = n\; . \tag{8.3b}$$

Let $\quad C := C_1 \,.\, \bar{C}\,$.

Then any desired pole configuration of the closed loop system can be achieved by choosing a suitable state feedback law. Especially the closed loop system can be designed to react as rapidly as desired to external disturbances or to input commands if the matrices Q and R are chosen properly.

This theoretical aspect of the linear optimization problem can be mathematically proved by applying techniques of the classical calculus of variations. The formal results may be stated as follows [8.1,8.2,8.3] :

Theorem 8.1:

Given the linear time invariant system (8.1) with optimization criterion (8.2a) and with $u(t)$ as a sufficiently smooth control vector. Assume the conditions (8.3a) and (8.3b) hold. Then we have :

(a) The performance index (8.2a)associated with (8.1) will be minimized by a control vector

$$u^o(t) = - R^{-1}.B^T.p(t) \quad , \tag{8.4}$$

where $p(t)$ is the adjoint vector of the state vector $x(t)$ (sometimes called Lagrangian Multiplicator). Both, $x(t)$ and $p(t)$, satisfy the related equations of Euler and Lagrange

$$\begin{pmatrix} \dot{x} \\ \dot{p} \end{pmatrix} = \begin{pmatrix} A & , -B.R^{-1}B^T \\ -\bar{C}^T.Q.\bar{C} , & -A^T \end{pmatrix} . \begin{pmatrix} x \\ p \end{pmatrix} ; \begin{pmatrix} x_o \\ p(t_e) \end{pmatrix} = \begin{pmatrix} x(o) \\ o \end{pmatrix} . \tag{8.5}$$

(b) Equation (8.5) provides a linear boundary value problem with a system matrix E which, in general, is unstable:

$$E = \begin{pmatrix} A & ,-B.R^{-1}.B^T \\ -\bar{C}^T.Q.\bar{C} , & -A^T \end{pmatrix} \text{ or } E = \begin{pmatrix} A & , -B.R^{-1}. B^T \\ -C^T.C, & -A^T \end{pmatrix} . \tag{8.6}$$

(c) The adjoint vector $p(t)$ may be derived from the state vector $x(t)$ by means of the relation

$$p(t) = P(t,t_e).x(t) \quad , \tag{8.7}$$

where $P(t,t_e)$ is solution of the Matrix-Riccati-Equation

$$\dot{P}(t) =-P(t).A-A^T.P(t)+P(t).B.R^{-1}.B^T.P(t)-\bar{C}^T.Q.\bar{C}.P(t) . \tag{8.8}$$

Then the optimal feed-back-law takes the form

$$u^o(t)= -K.x^o(t) \quad , \text{ where } \quad K=R^{-1}.B^T.P(t,t_e) . \tag{8.9}$$

(d) We have

$$\lim_{t_e \to \infty} P(t,t_e) = P \quad , \qquad (8.10)$$

where the constant matrix P is solution of the stationary Riccati - Equation for $\dot{P}(t) = 0$.

(e) The optimal stationary system has the system matrix

$$A^o = A - B.R^{-1}.B^T.p \quad . \qquad (8.11)$$

It is asymptotically stable.

(f) The eigenvalues of the matrix A^o are identical to those eigenvalues of the matrix E with negative real parts.

The direct solution of the above optimization problem by on - line - simulation of equations (8.5) together with (8.4) leads to a time invariant dynamic system which may be interpreted very easily by means of frequency domain techniques. Unfortunately, it is a boundary value problem with an unstable system matrix E and with an, in general, unknown boundary value $p(t_e)$. Therefore, the vector $p(t)$ is usually eliminated by means of the relations (8.4), (8.7) and (8.8) to yield the linear control law (8.9). As a consequence, the nonlinear system of differential equations (8.8) must only be solved once (off-line). Then the related control law can be storaged in an on-line-controller. Unfortunately, this control law is time variant for finite values of the parameter t_e . Therefore, the final time t_e is usually chosen infinite in order to get a time invariant linear optimal control law.

From the point of view of practice, the choice of the matrices Q and R in (8.2a) and (8.2b) turns out to be one of the essential difficulties . Here, practical experience and intuition play a major role (comparable to the trial and error methods of classical control technique).

Based on these considerations, the design of optimal control systems has been combined with the classical root locus technique in [8.3] and [8.4] .
These investigations yield the following relations between a simplified version of the optimization criterion (8.2a) and the position of the closed loop poles of the optimal control system.
Starting from relation

$$x_o = B.z_o \quad , \quad \text{where} \quad z(t) = z_o.\phi(t) \quad , \tag{8.12a}$$

the transfer function matrix $F^o(s)$ of the optimal control system as reaction to the input $z(s)$,i.e. $y(s)=F^o(s).z(s)$, may be written in the form [8.3]

$$F^o(s)=F(s).\{R+F^T(-s).Q.F(s)\}^{-1}.\{R+B^T.\phi^T(-s).P(o,t_e).B\}, \tag{8.12b}$$

where

$$F(s) = \bar{C}.(sI_n - A)^{-1}.B \tag{8.13a}$$

is the transfer matrix of the control plant, $\phi(s)$ is defined as

$$\phi(s) := (sI_n - A)^{-1} \tag{8.13b}$$

and $\det \phi^{-1}(s)$ is the characteristic equation of the control plant

$$\det \phi(s)^{-1} = \det (s.I_n - A) \quad . \tag{8.13c}$$

Using relation (8.9) yields on the other hand ,

$$F^o(s) = F(s).V^{-1}(s) \tag{8.14}$$

where

$$V(s) = (I_q + K.\ \phi(s).B) \tag{8.15}$$

is the return difference of the control loop and $\det (V(s))$ is the dynamic control factor.

According to statement (e) of Theorem 1, the relation

$$\det\ (V(s) = \det\ (I_q + K.\ \phi(s).B) \qquad (8.16)$$

is a Hurwitz-Polynomial.

Applying some elementary manipulations [8.3] , the relation between the matrices Q and R and the poles of the optimal control system for $t_e = \infty$ may be stated in the form

$$E(s) = V^T(-s).R.V(s) = R + F^T(-s).Q.F(s) \qquad (8.17)$$

According to the previous results (Theorem 1.e), the optimal linear control strategy yields for $t_e = \infty$ a time-invariant feedback law. According to [8.5] , the zeros of a linear control system (in the sense of Mc Millan) are invariant under state feedback. Therefore, the above task is solved if the formal relation

$$\det\ (E(s)) = 0 \qquad (8.18)$$

between the closed loop poles and the matrices R and Q has been investigated.

Unfortunately, the geometrical representation of this relation in terms of a root locus plot only admitts the variation of one scalar parameter or the variation of the 2.n.n elements of the matrices Q and R, each proportional to the others.

Therefore the relation (8.2a) is specialized as follows

$$J = \int_0^\infty \{y^T.Q.\, y + \varrho.u^T.R.u\}\, dt \quad . \qquad (8.2')$$

Relation (8.19) in connection with the higher order root locus technique, applied to (8.18), provides the compromise wanted between reaction velocity of the system and the amound of control energy needed to achieve this reaction velocity. Using $k := 1/\rho$ and (8.2') in connection with the previous relations, equation (8.18) yields [8.3] :

$$\det(s.I_n - E) = \det \begin{pmatrix} sI - A & +k.B.R^{-1} \\ \bar{C}^T.Q.\bar{C} & sI+A^T \end{pmatrix} . \quad (8.19)$$

Using the determinant identities

$$\det\begin{pmatrix} A_{11} & A_{12} \\ -A_{21} & A_{22} \end{pmatrix} = \det A_{11}.\det(A_{21}.A_{11}^{-1}.A_{12}+A_{22}), \det A_{11} \neq 0 \quad (8.20)$$

and

$$\det(I_n + AB) = \det(I_n + BA) \quad , \quad A,B \in \mathbb{R}^{n,n} \quad (8.21)$$

yields the relation

$$\det(sI_n - E) = (-1)^n . \psi_A(s) . \psi_A(-s).\det\{I+k.R^{-1}.F^T(s).Q.F(s)\} \quad (8.22)$$

where

$$\psi_A(s) := \det(s.I_n - A) \quad . \quad (8.23)$$

This relation has the form of equation (7.10) of Chapter VII. It can therefore be investigated on the basis of the <u>higher order root locus technique</u>.
Applying this technique to the above problem, the following statement holds :

<u>Theorem 8.2</u>:

Given a completely controllable and completely observable control plant with p inputs u_i and p outputs y_j . Assume the plant has a regular transfer function matrix F(s) (with $\det\{F(s)\} \neq 0$). On the basis of the McMillan Normal Form of F(s) we have

$$\det F(s) = \frac{Z(s)}{\psi_A(s)} = k_1 . \frac{\prod_{i=1}^{m}(s+s_i)}{\prod_{j=1}^{n}(s+p_j)} \quad , \quad m \leq n \; , \quad (8.24)$$

where s_i and p_j are the zeros and poles of F(s) in the sense of McMillan, respectively.

According to the higher order root locus construction rules, the closed loop poles of the optimal system transfer function matrix $F^{o}(s;k)$ change with k according to the following statements :

(a) For increasing $k(k\rightarrow\infty)$, exactly m poles of the optimal closed loop system transfer function $F^{o}(s,k)$ tend towards the points

$$\hat{s}_i = \begin{cases} s_i & \text{for} \quad \mathrm{Re}\{s_i\} \leq 0 \\ -s_i & \text{for} \quad \mathrm{Re}\{s_i\} > 0 \end{cases} \quad , i = 1, \ldots , m \; .$$

(b) The residual (n-m) poles p_j tend for $k\rightarrow\infty$ towards infinity, where three cases may occur :

Case 1 :

Assume, a straight line of segment VIII meets exactly two points of the exponent diagram, and there is no point of the exponent diagram inside the area limited by this segment line and by the related perturbation line.
Then for $k\rightarrow\infty$, the related root locus branches tend towards asymptotes which constitute a Butterworth Configuration. They are easily computed according to Rules 14 and 15 of Chapter IV.

Case 2 :

Assume, a straight line of segment VIII meets more than two points of the exponent diagram, and there is no point of the exponent diagram inside the area limited by this segment line and by the related perturbation line.
Then for $k\rightarrow\infty$, the related root locus branches tend towards asymptotes which don't constitute a Butterworth Configuration. They are computed according to Rules 14 and 15 of Chapter IV.

Case 3 :

Assume there are points of the exponent diagram inside the area limited by a segment line and by the related perturbation line of Segment VIII.
Then the related root locus branches don't tend towards asymptotes with finite asymptote points.

(c) For $k=o$, the n root locus branches of the closed loop transfer function matrix $F^{o}(s,k)$ start at points

$$\hat{p}_j := \begin{cases} p_j & \text{for} \quad Re\{p_j\} \leq 0 \\ -p_j & \text{for} \quad Re\{p_j\} > 0 \end{cases} \qquad , j = 1, \ldots , n$$

(compare Rules 1o and 11 of Chapter IV).

Note :

(i) Cases (a) and (b) of Theorem 8.2 ($k \to \infty$) accent the reaction velocity of the closed loop system compared to the demand of minimal control energy.

(ii) Case (c) of Theorem 8.2 lays stress upon minimal control energy of the closed loop system.

(iii) In case of a singular open loop transfer function matrix $F(s;k)$, the statements of Theorem 8.2 hold if the concept of the McMillan zeros is replaced by the concept of the sinks of the system (compare Chapter VII).

The above discussion will now be interpreted by an example. The system (8.1) and the optimization criterion (8.2a) are fixed by the matrices

$$A = \begin{pmatrix} -1 & 1 \\ 0 & -2 \end{pmatrix}, \quad \bar{C} = \begin{pmatrix} 1 & 0 \\ 0 & 1 \end{pmatrix}, \quad B = \begin{pmatrix} 1 & 0 \\ 1 & 1 \end{pmatrix} \qquad (8.25)$$

and

$$R = Q = I_2 = \begin{pmatrix} 1 & 0 \\ 0 & 1 \end{pmatrix} . \tag{8.26}$$

Then, according to (8.13a) and to (8.13c), we have

$$F(s) = \frac{1}{(s+1).(s+2)} \cdot \begin{pmatrix} s+3 , & 1 \\ s+1 , & s+1 \end{pmatrix} \tag{8.27}$$

and

$$\Psi_A(s) = \det(sI - A) = (s+1).(s+2) \tag{8.28}$$

or

$$\Psi_A(s) \,.\, \Psi_A(-s) = s^4 - 5.s^2 + 4 \tag{8.29}$$

and

$$F^T(-s).F(s) = (\Psi_A(s).\Psi_A(-s))^{-1} . \begin{pmatrix} 10 - 2.s^2 , & 4-s-s^2 \\ 4+s-s^2 , & 2-s^2 \end{pmatrix} . \tag{8.30}$$

The characteristic polynomial of the Euler-Lagrangian-Equation takes the form

$$\Psi_E(s) = \det(s.I_4 - E) = (\Psi_A(s).\Psi_A(-s))^{-1}.\det(I_4 + k.F^T(-s).F^T(s)) \tag{8.31}$$

or

$$\Psi_E(s) = (\Psi_A(s).\Psi_A(-s))^{-1}.\det \begin{pmatrix} (s^4 - 5.s^2 + 4) + (10 - 2.s^2).k ; & k.(4-s-s^2) \\ k.(4+s-s^2) & ; (s^4 - 5.s^2 + 4) + (2-s^2).k \end{pmatrix} ,$$

and finally

$$\Psi(s) = (4 - 5.s^2 + s^4) + k.(12 - 3.s^2) + k^2 \tag{8.32}$$

or

$$\Psi(s) = (s+1).(s+2).(-s+1).(-s+2) + 3.k.(s-2).(-s-2) + k^2 . \tag{8.33}$$

The expression $F^T(-s).F^T(s)$ has the McMillan normal form

$$M_C(s) = \begin{pmatrix} (\psi_A(s).\psi_A(-s))^{-1} & , & 0 \\ 0 & , & 1 \end{pmatrix} . \tag{8.34}$$

Relations (8.33) and (8.34) show that neither the control plant nor the optimal closed loop control system have finite zeros.

The root locus plot corresponding to the Euler-Lagrangian-Equation will be computed according to the higher order root locus construction rules.

Construction of the root locus plot

Rule 1 : Polynomial coefficients of $Q_i(s)$ and $R_i(k)$.

	s^0	s^1	s^2	s^3	s^4	
Q_2	1					k^2
Q_1	12	o	-3			k^1
Q_0	4	o	-5	o	1	k^0
	R_0	R_1	R_2	R_3	R_4	

Table 8.a.1 : Polynomial coefficients of $Q_i(s)$ and $R_j(k)$.

Rule 2 : Roots of polynomials $Q_i(s)$ and $R_j(k)$

Q_i	i	$s-s_1^i$	$s-s_2^i$	$s-s_3^i$	$s-s_4^i$	
Q_1	1	(s+2)	(s-2)			k^1
Q_0	o	(s+1)	(s+2)	(s-1)	(s-2)	k^0
		R_0	R_1	R_2	R_3	

Table 8.b.1 : Roots of $Q_i(s)$

	s^0	s^1	
Q_1	k+11,66		$k-k_j^1$
Q_0	k+o,34	k+1,66	$k-k_j^0$
j	o	1	
R_j	R_0	R_1	

Table 8.c.1 : Roots of $R_j(k)$

Rule 3 and 4 : Exponent Diagram

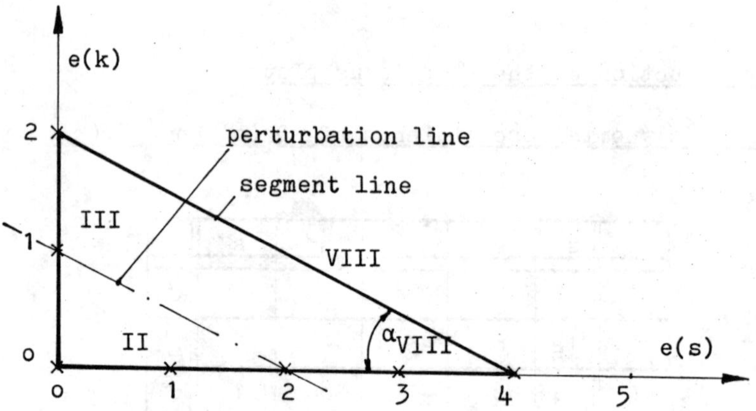

Figure 8.1.a : Exponent Diagram

Rule 5 : Constants

$l_0 = 0$, $q = 2$, $n = 4$, $n_0 = 4$

$e_I = 0$, $e_{II} = 4$, $e_{IV} = 0$, $e_V = 0$

$e_{VII} = 0$, $e_{VIII} = 4$, $\bar{e}_{III} = 2$, $\bar{e}_{VI} = 0$

Rule 6 : Slopes

$$\beta_{VIII} = 1/2$$

Note :

There exist $n = n_o = 4$ root locus branches which start at the poles $s_1^o = -1$, $s_2^o = -2$, $s_3^o = +1$ and $s_4^o = +2$ and which tend to infinity for increasing $k, k > o$. No branch runs through the point $s = o$.

Rule 11 : Angles of departure from the points s_j^o (j=1,2,3,4); (segment II, k = o).

Replacing the variable s of $Q_o(s)$ and $Q_1(s)$ by the variables $s'-1$, $s'-2$, $s'+1$ and $s'+2$ yields for

$s_1^o = -1$:

$$Q_o(s') = o + 6.s' + \dots ,$$
$$Q_1(s') = 9 + \dots ,$$
$$\gamma = -1{,}5 , \quad \beta_I = 1 \quad \text{and}$$
$$\varphi(-1{,}o) = \pi \quad \text{for} \quad k > o ;$$

for

$s_2^o = +1$:

$$Q_o(s') = o - 6.s' + \dots ,$$
$$Q_1(s') = 9 + \dots ,$$
$$\gamma = 1{,}5 , \quad \beta_I = 1 \quad \text{and}$$
$$\varphi(1{,}o) = o \quad \text{for} \quad k > o ;$$

for

$s_3^o = -2$:

$$Q_o(s') = o - 12.s' + \dots ,$$
$$Q_1(s') = o + \dots ,$$
$$Q_2(s') = 1 ,$$
$$\gamma = 1/12 > o , \quad \beta_I = 2 \quad \text{and}$$

$\varphi(-2,0) = 0$ for $k > 0$ and for $k < 0$;

for

<u>$s_4^o = +2$:</u>

$Q_0(s') = 0 + 12.s' + \ldots$,

$Q_1(s') = 0 + \ldots$,

$Q_2(s') = 1$,

$\gamma = -1/12 < 0$, $\beta_I = 2$ and

$\varphi(2,0) = \pi$ for $k > 0$ and for $k < 0$.

<u>Rule 14 :</u> <u>Asymptote angles (segment VIII, $k = \infty$).</u>

Using $\beta_{VIII} = 2/4 = 0,5$,

the related supporting polynomial

$$\gamma^4 - 3.\gamma^2 + 1 = 0$$

with roots

$$\gamma_1 = +\sqrt{2,62} \quad , \quad \gamma_2 = -\sqrt{2,62}$$
$$\gamma_3 = +\sqrt{o,38} \quad , \quad \gamma_4 = -\sqrt{o,38}$$

or

$$\arg\gamma_1 = 0$$
$$\arg\gamma_2 = \pi$$
$$\arg\gamma_3 = 0$$
$$\arg\gamma_4 = \pi$$

yields the asymptote angles

$$\left.\begin{array}{l}\varphi_1(\infty,\infty) = \varphi_3(\infty,\infty) = 0 \\ \varphi_2(\infty,\infty) = \varphi_4(\infty,\infty) = o\end{array}\right\} \text{ for } k > 0$$

and

$$\varphi_1(\infty,\infty) = \varphi_3(\infty,\infty) = \frac{\pi}{2}$$

$$\varphi_2(\infty,\infty) = \varphi_4(\infty,\infty) = \frac{3}{2}\cdot\pi \cong -\frac{\pi}{2} \qquad \text{for} \quad k < 0 \quad .$$

<u>Note :</u>

Because of the results of Rule 14, the asymptote points have not to be computed.

<u>Rule 17 :</u> <u>Departure of root locus branches from $s_j^o = o$ (segment III, $k = k_o^j \neq o$)</u>

Because of Table 8.c.1, no branch meets the point s=o for k≥o.

<u>Note :</u>

Because of the relation

$$n = n_o = n_i + 2 \quad , \quad \text{for} \quad i > o$$

the sum of the real parts of all roots of (8.32) is constant (compare (4.44) in Step 21 of Chapter IV).

The root locus plots of (8.32) are drawn in Figures (8.1.b) and (8.1.c).
These figures show that the asymptotes don't constitute a Butterworth Configuration (in agreement with Theorem 2b, Case 2, but contrary to the results in [8.3] and [8.4]).

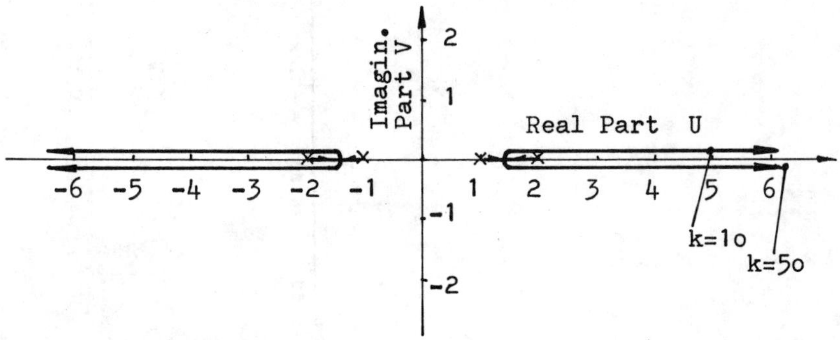

Figure 8.1.b: Qualitative root locus plot .

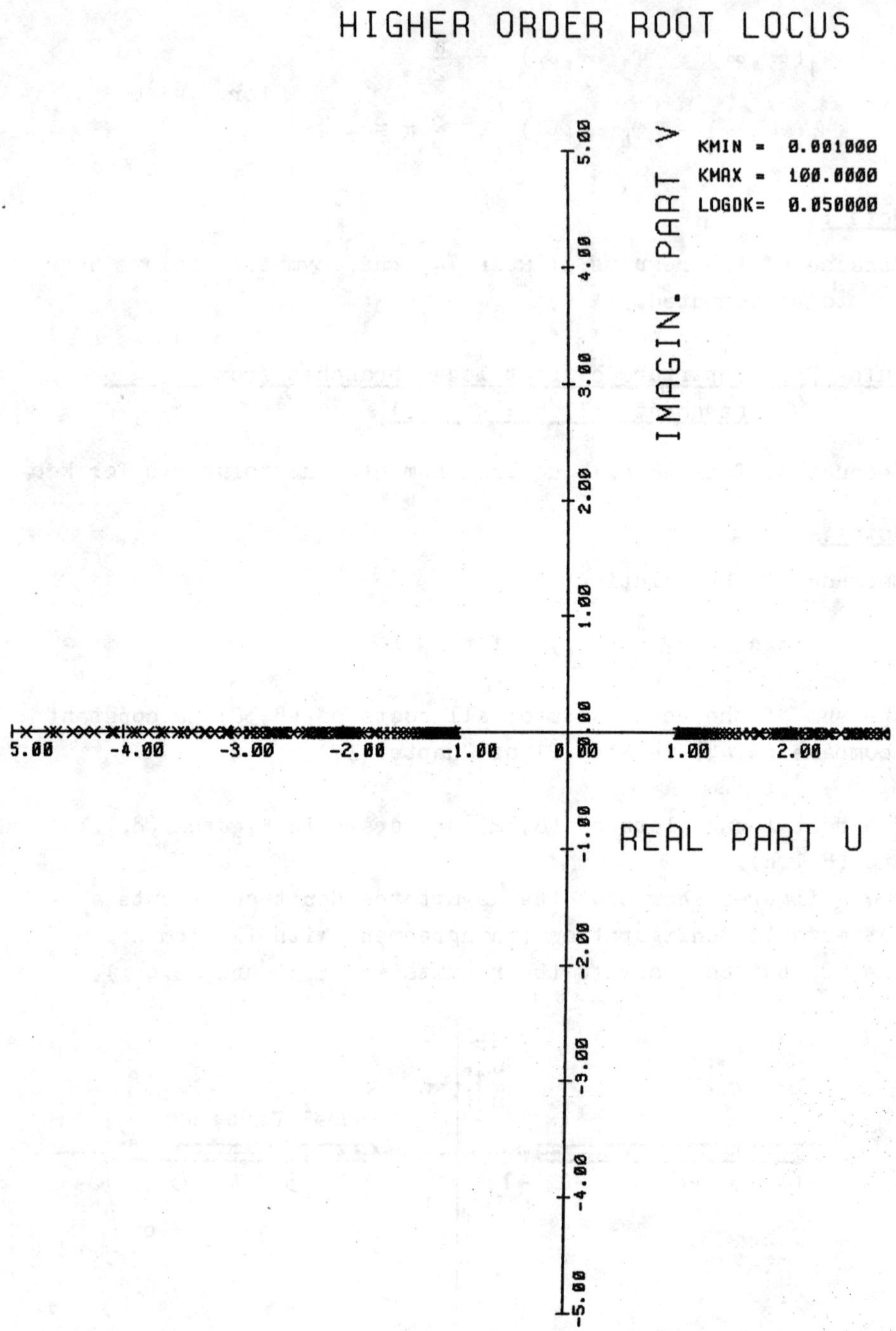

Figure 8.1.c: Computer root locus plot .

References

[8.1] R.E. Kalman, Contributions to the Theory of Optimal Control, Bol. Soc. Mat. Mex., 5 , 1o2-119, (196o).

[8.2] E.Lee, L.Marcus, Foundations of Optimal Control Theory, J. Wiley and Sons, 1967.

[8.3] H. Schwarz, Optimale Regelung linearer Systeme, BI-Wissenschaftsverlag, 1976.

[8.4] K. Heym, Ein Beitrag zur Theorie des strukturoptimalen Regelungssystems, Diss. TU-Hannover, 1972.

[8.5] B. McMillan, Introduction to Formal Realizibility Theory, The Bell System Technical Journal, Vol. XXXI, No.2, 1952, pp.217-279.

Appendix A: The Formal Basis of Newton's Diagram Technique

This appendix yields the formal basis of the Newton Diagram Technique which, in its turn, delivers the fundamental tool of the Exponent Diagram Technique and of the higher order root locus technique, developped by the author in chapter III and chapter VI of this monograph. Different versions of a formal proof of the Newton Diagram Technique may be found in [A.1] and [A.2]. The proof of this technique has been reformulated here from a more geometric point of view, taking into consideration some notions which will turn out to be very useful in connection with the construction and interpretation of the higher order root locus rules. Two different situations are treated here :

A.1 : The case of formal power series and
A.2 : The case of analytic functions.

A.1 Formal Power Series

Let $\overline{F} : U_1 \times U_1 \longrightarrow \mathbb{C}^1$, where $\overline{F} \in I_2$ and $I_2 := K^*\{x,k\}$ is the ring of the formal power series in two variables (x and k), $x, k \in \mathbb{C}^1$ (complex numbers),
$U_1 \times U_1 \subset \mathbb{C}^1 \times \mathbb{C}^1$ and $(x,k) \in U_1 \times U_2$.
Then $\overline{F}(x,k)$ may be written in the form

$$\overline{F}(x,k) = \sum_{i=0}^{\infty} \overline{F}_i(k).x^i = \sum_{i,j=0}^{\infty} \overline{F}_{ij}.x^i.k^j \qquad (A.1)$$

, where $\overline{F}_i(k) \in I_1$ and $I_1 := K^*\{k\}$ is the ring of the formal power series in k.
A root x(k) of (A.1) is called small solution of (A.2) iff the following relation is fulfilled

$$\lim_{k \to 0} x(k) = 0 .$$

We are interested in all of the small solutions of equation (A.2) near the point $(x,k) = (o,o)$, especially in case of the critical situation (A.3) :

$$\overline{F}\,(x,k) = o \quad , \qquad (A.2)$$

$$\partial\overline{F}/\partial x(o,o) = o \quad . \qquad (A.3)$$

In view of the fact that even the very simple algebraic equation

$$x^2 - k = o \qquad (A.4)$$

has solutions of the form

$$x_1 = k^{1/2} \quad \text{and} \quad x_2 = -\,k^{1/2} \quad , \qquad (A.5)$$

it is useful to investigate the small solutions of equation (A.6)

$$F(x,k) = \sum_{i=o}^{\infty} F_i(k).x^i = \sum_{i,j=o}^{\infty} \overline{F}_{ij}.x^i.k^{\frac{j}{r}} = o \quad , \qquad (A.6)$$

where $F_i(k) \in K^r\{k\}$ and $K^r\{k\}$ is the ring of the formal power series of the form

$$x(k) = \sum_{j=o}^{\infty} \gamma_j \,.\, \lambda^{\frac{j}{r}} \;,\; o<r \in \mathbb{N} \;,\; j \in \mathbb{N} \quad . \qquad (A.7)$$

Let $K^{\infty}\{k\} := \bigcup_{r \in \mathbb{N}} K^r\{k\}$.

Then $K^r\{k\}$ and $K^{\infty}\{k\}$ are subrings of $K^{*}\{k\}$.

Expressing $F_i(k)$ in the form

$$F_i(k) = k^{\varrho_i} . \sum_{j=0}^{\infty} F_{ij} . k^{j/r} \quad , \tag{A.8a}$$

where

$$F_{ij} = \overline{F}_{i,j+\varrho_i} \; ; \varrho_i := d_i/r \geq 0 \; ; \; F_{i,0} \neq 0 \; ; \; d_i, r \in \mathbb{N} \; ; \; r > 0, \tag{A.8b}$$

we have

$$F(x,k) = \sum_{i=0}^{\infty} k^{\varrho_i} . x^i . \sum_{j=0}^{\infty} F_{ij} . k^{j/r} = 0 \, . \tag{A.9}$$

Before formulating the basic theorems concerning the existence and the number of all small solutions of equation (A.9) near the point $(x,k) = (0,0)$, the following notation and definitions are in troduced :

DEFINITION A.1 :

Starting from the equation (A.9), it is useful to define:

a. $h(F) := \min_i(\varrho_i)$ is called <u>height of F</u> .

b. $s(F) := i'$ is called <u>degree of F</u> (with respect to x), where i' is the index i of ϱ_i corresponding to h(F).

c. Any pair of exponents (i, ϱ_i) of equation (A.9) may be interpreted as a point within the i-ϱ_i-plane. The descending part of the convex polygon envelopping the set of points (i, ϱ_i) from below is called <u>NEWTON POLYGON</u> (compare Fig. A.1). We have $0 \leq i \leq s(F)$ for all points of equation (A.9) which together with the coordinates yields the <u>NEWTON DIAGRAM</u> .

d. $\mathcal{L}_j$ is one of the different straight lines of $\mathcal{L}$. We have $\mathcal{L} = \bigcup_j \mathcal{L}_j$.

e. $l(\mathcal{L})$ is the <u>length of $\mathcal{L}$</u>. It is defined as the length of the projection of the Newton Polygon onto the i-axis. We have $0 \leq l(\mathcal{L}) \leq s(F)$.

f. $l_0(\mathcal{L})$ is called the <u>degree of degeneration of $\mathcal{L}$</u>. It is defined by means of the relation

$$l_0(\mathcal{L}) := s(F) - l(\mathcal{L}) .$$

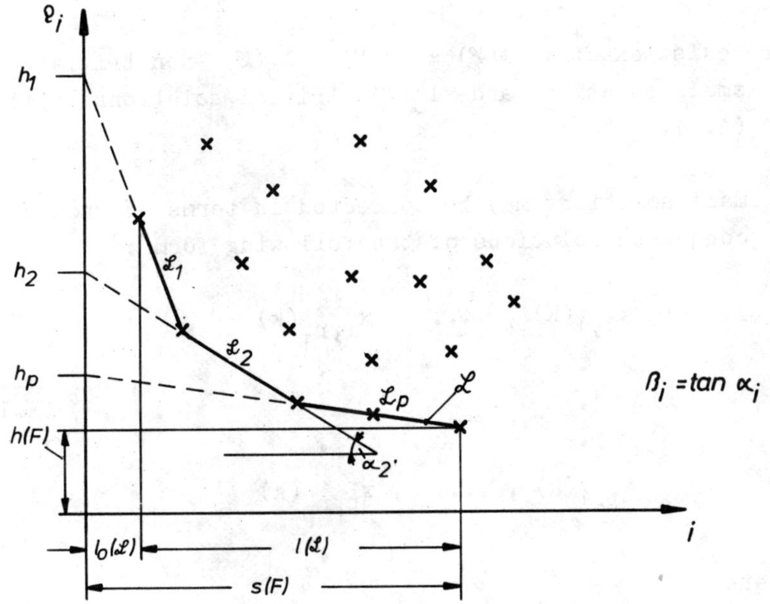

Fig. A.1 : Newton Diagram of the relation $F(x,k) = 0$.

The basic results in connection with the computation of all small solutions of the equations (A.6) and (A.9) are summarized in theorem A.1 .

THEOREM A.1 :

Let $h(F) = 0$. Then the following statements hold :

a. There exists at least one small solution $x(k) \in K^*(k)$ of equation (A.9) iff $s(F) > 0$.

b. There exist exactly $s(F)$ small solutions $x(k) \in K^*(k)$ of equation (A.9) (some of which may be real or complex, simple or multiple, trivial or non trivial).

c. There exist exactly $l(\mathcal{L}) = s(F) - l_o(\mathcal{L})$ non trivial small solutions and $l_o(\mathcal{L})$ trivial solutions $x(k)$ of (A.9).

d. The small solutions may be collected in terms of groups of conjugate solutions of the following form :

$$\begin{matrix} x_{1,1}(k) , & \dots , & x_{1,r_1}(k) & , \\ \cdot & & \cdot & \\ \cdot & & \cdot & \\ \cdot & & \cdot & \\ x_{p,1}(k) , & \dots , & x_{p,r_p}(k) & , \end{matrix} \qquad \text{(A.10)}$$

where

$$\sum_{j=1}^{p} r_j = l(\mathcal{L}) .$$

The statements of theorem A.1 will be proved in different steps which are formulated in terms of some lemmata. Within this context,the following expressions will be used.Let

$$\varrho_i = h_j - i.\beta_j \quad , \quad \beta_j := \tan \alpha_j \, , \; j=1,\dots,p \qquad \text{(A.11)}$$

be the equation which describes the straight line $\mathscr{L}_j$ of the Newton Polygon with slope α_j (compare Fig. A.1). Let

$$\Psi_j(\gamma) := \sum_{\{i:h_j=\varrho_i+i.\beta_j\}} F_{i,o}.\gamma^i \tag{A.12}$$

be the <u>supporting polynomial</u> of $\mathscr{L}_j$ of degree $l(\mathscr{L}_j)$ with respect to γ. It contains all terms of the equation (A.9) the corresponding points of which are met by the straight line $\mathscr{L}_j$. Then the following lemmata may be formulated:

LEMMA A.2 :

Let $l(\mathscr{L}) > o$. Then the following statements hold :

a. The constants β_j, $h_j \in \mathbb{Q}$ (rational numbers). They satisfy the relation

$$h_j - h(F) \geq \beta_j \tag{A.13}$$

, where

$$\beta_j = \frac{p_j}{r.q_j} \quad ; \; p_j, q_j, r, j \in \mathbb{N} \quad .$$

b. Let p_j and q_j of (A.13) be relative prime. Then q_j divides the constant $l(\mathscr{L}_j)$, and we have

$$\Psi_j(\gamma) = \gamma^{\alpha}.\phi_j(\gamma^{q_j}) \quad , \; \alpha \in \mathbb{N} \tag{A.14}$$

,where ϕ_j is a polynomial of degree $l(\mathscr{L}_j)/q_j$ in its argument.

Proof :

a. Let the pairs $(\lambda, \varrho_\lambda)$ and $(\varkappa, \varrho_\varkappa)$ be the boundary points of the straight line $\mathscr{L}_j$ for $\lambda < \varkappa$. Then from (A.11) we have

$$\varrho_\lambda + \lambda.\beta_j = h_j = \varrho_\varkappa + \varkappa.\beta_j$$

or

$$\beta_j = \frac{\varrho_\lambda - \varrho_\varkappa}{\varkappa - \lambda}, \text{ and using the relation } \varrho_i := d_i/r$$

yields

$$\beta_j = \frac{d_\lambda - d_\varkappa}{r.(\varkappa-\lambda)} \quad ; \quad d_i \; , \; r \; , \; \lambda \; , \; \varkappa \in \mathbb{N} \quad , \tag{A.15}$$

where $\beta_j, h_j \in \mathbb{Q}$, $r.(\varkappa - \lambda) = l(\mathscr{L}_j)$ and β_j may be written in the form

$$\beta_j = \frac{p_j}{q_j.r} \quad ,$$

where p_j and q_j are relative prime (after cancellation of the greatest common divisor (g.c.d.) $\bar{\mu}$ of the terms $(d_\lambda - d_\varkappa)$ and $(\varkappa - \lambda)$). For $\bar{\mu} > 1$, the constant q_j is a nontrivial divisor of the factor $(\varkappa - \lambda) = l(\mathscr{L}_j)$.
Combining the relations $h(F) := \min\limits_i \varrho_i$ and (A.15) yields the relations

$$\varrho_\lambda \geq h(F) \; , \quad \lambda . \beta_j = h_j - \varrho_\lambda \leq h_j - h(F) \tag{A.16}$$

and

$$\beta_j \leq \frac{1}{\lambda}.(h_j - h(F)) \qquad \text{or for } \lambda \geq 1$$

$$\beta_j \leq (h_j - h(F)) \quad . \tag{A.13}$$

b. Let (i, ϱ_i) be any point on the straight line $\mathscr{L}_j$ of the Newton Polygon. Then we have

$$\beta_j = \frac{p_j}{q_j.r} = \frac{\varrho_\lambda - \varrho_i}{(i - \lambda)} = \frac{d_\lambda - d_i}{r.(i-\lambda)} \quad , \quad (d_i = r.\varrho_i)$$

or

$$p_j.(i - \lambda) = q_j.(d_\lambda - d_i) \quad . \tag{A.17}$$

Therefore q_j not only divides the factor $l(\mathscr{L}_j) = (\varkappa - \lambda)$ but also the factor $(i - \lambda)$ for $\lambda \leq i \leq \varkappa$.

Then we have

$$i = \lambda + \mu . q_j \quad \text{or} \quad i - \lambda = \mu . q_j \ , \ \mu \in \mathbb{N}. \tag{A.18}$$

Using these expressions, the supporting polynomial $\Psi_j(\gamma)$ of $\mathcal{L}_j$ (A.12) takes the form

$$\Psi_j(\gamma) = \sum_{\{i:\varrho_i+i.\beta_j=h_j,\lambda\le i\le\varkappa\}} F_{io}.\gamma^i \quad \text{or}$$

$$\Psi_j(\gamma) = \gamma^\lambda . \sum_{\{i:\varrho_i+i.\beta_j=h_j,\lambda\le i\le\varkappa\}} F_{io}.\gamma^{i-\lambda} \ , \tag{A.19}$$

and in connection with (A.18)

$$\Psi_j(\gamma) = \gamma^\lambda . \sum_{\{i:\varrho_i+i.\beta_j=h_j\}} F_{io}.(\gamma^{q_j})^{(i-\lambda)/q_j} \tag{A.20a}$$

or

$$\Psi_j(\gamma) = \gamma^\lambda . \sum_{\mu=0}^{(\varkappa-\lambda)/q_j=l(\mathcal{L}_j)/q_j} \tilde{b}_\varkappa.(\gamma^{q_j})^\mu =: \gamma^\lambda . \phi_j(\gamma^{q_j}) \ , \tag{A.20b}$$

where

$$\tilde{b}_\varkappa = \begin{cases} F_{\lambda+\mu.q_j,o} & \text{for } \varrho_{\lambda+\mu.q_j} + (\lambda+\mu.q_j).\beta_j = h_j \\ & \text{(points on } \mathcal{L}_j) \\ 0 & \text{for } \varrho_{\lambda+\mu.q_j} + (\lambda+\mu.q_j).\beta_j > h_j \\ & \text{(points beyond } \mathcal{L}_j) \end{cases} \tag{A.20c}$$

and (i,ϱ_i) defines a point on $\mathcal{L}_j$, where

$0 \le \mu \le l(\mathcal{L}_j)/q_j$ for $\lambda \le i \le \varkappa$, and $l(\mathcal{L}_j)/q_j$ is the degree of the polynomial $\phi_j(\gamma)$. ∎

The following lemma yields a procedure to construct all small solutions of (A.9).

LEMMA A.3 :

a. The equation (A.9) has exactly $l_o(\mathcal{L})$ trivial solutions running through the point $(x,k) = (o,o)$.

b. The equation (A.9) has exactly $l(\mathcal{L})$ non trivial solutions running through the point $(x,k) = (o,o)$.

c. All non trivial small solutions of (A.9) running through the point $(x,k) = (o,o)$ are of the form

$$x(k) = \gamma_1 . k^{\beta_1} + o(|k|^{\beta_1}) \quad , \tag{A.21}$$

where $\beta_1 = \tan\alpha_1$ and α_1 is the slope of one of the straight lines $\mathcal{L}_j$ of $\mathcal{L}$, and γ_1 is one of the non vanishing roots of the supporting polynomial $\psi(\gamma_1)$ corresponding to $\mathcal{L}_j$.

Proof :

a. Let $l_o(\mathcal{L}) > o$. Then in all of the terms of the series (A.9), the factor x has an exponent i bigger than or equal to $l_o(\mathcal{L})$. Therefore, from equation (A.9) a factor

$$x^{l_o(\mathcal{L})} = x^{(s(F)-l(\mathcal{L}))}$$

may be factored out. On the other hand, if $l_o(\mathcal{L}) = o$ then there exists a term $F_{ij}.k^{\varkappa}$ $(\varkappa>o)$ of (A.9), and no factor x^{μ} $(\mu>o)$ may be factored out.
As a result , equation (A.9) has exactly $l_o(\mathcal{L})$ small solutions of the form $x(k) \equiv o$ for all real values of k.

b. and c. Let x(k) be a small solution of (A.9) of the form (A.21), where $\beta_1 > o$ and $\gamma_1 \neq o$. Then $x(k) \in K^*\{k\}$. Inserting (A.21) into (A.9) yields the relation

$$F(x(k),k) = \sum_{i=o}^{\infty} F_{i,o}.\gamma_1^i.k^{\varrho_i}.k^{i.\beta_1} + o(|k^{\varrho_i}.k^{i.\beta_1}|) \tag{A.22a}$$

$$= \sum_{i=o}^{\infty} F_{i,o}.\gamma_1^i.k^{\varrho_i+i.\beta_1} + o(|k^{\varrho_i+i.\beta_1}|) \quad ,$$

and using the variable

$$h_1 := \min_i (\varrho_i + i.ß_1) \qquad (A.22b)$$

we have

$$\psi(\gamma_1) := \sum_{\{i:\ \varrho_i + i.ß_1 = h_1\}} F_{i,o}.\gamma_1^i \qquad (A.22c)$$

and

$$F(x(k),k) = \psi(\gamma_1).k^{h_1} + o(|k|^{h_1}) \ . \qquad (A.22d)$$

By assumption, $x(k)$ is a solution of the equation (A.9). Therefore, $F(x(k),k) \equiv 0$ for all real values of k. Then $\psi(\gamma_1) \equiv 0$ identically in k, and γ_1 is a non vanishing solution of $\psi(\gamma) = 0$. The polynomial $\psi(\gamma)$ has only a non vanishing solution $\gamma_1 \neq 0$ iff it includes at least two non vanishing terms, e.g. if the relation

$$\psi(\gamma_1) = F_{i_1,o}.\gamma_1^{i_1} + F_{i_2,o}.\gamma_1^{i_2} + \dots \qquad (A.23a)$$

holds for $F_{i_1,o} \neq 0$ and for $F_{i_2,o} \neq 0$.

Then the equation

$$h_1 = \varrho_i + i.ß_1 \qquad (A.23b)$$

has at least two real solutions i_1 and i_2 , and because of (A.22b) all admissable i which are in contradiction to (A.22c) satisfy the relation

$$h_1 < \varrho_i + i.ß_1 \quad . \qquad (A.23c)$$

This implies that the numbers h_1 and $ß_1$ in (A.21) are not independent of each other. They must satisfy the relations (A.22c) and (A.23b). The basic problem is to determine the coupled constants h_1 and $ß_1$. Replacing in (A.22c), the constants γ_1, h_1 and $ß_1$ by means of the constants γ, h_j and $ß_j$, the function ψ by means of ψ_j and the number 1 by means of the variable j yields the relations (A.12) and (A.11)(i.e. the defining equations of the straight line $\mathcal{L}_j$ and of the corresponding supporting polynomial ψ_j).

This important result allows to determine the constants β_1 and γ_1 of the solution (A.21) from the Newton Diagram corresponding to (A.9). Thus, the ansatz (A.21) satisfies the equation (A.9) iff γ_1 is a (non vanishing) root of the supporting polynomial ψ_j, and β_1 is the slope of the corresponding straight line $\mathscr{L}_j$ of the Newton Polygon. The existence of a non vanishing root γ_1 of (A.22c) is based on the assumption that there exist at least two non vanishing terms of (A.22c). This implies the relations

$$l(\mathscr{L}_j) > 0 \quad \text{and} \quad l(\mathscr{L}) > 0 \quad .$$

Lemma A.3 yields a graphical procedure to construct all small solutions of equation (A.9) (till now to first order of accuracy). The heuristic idea of this procedure as well as its practical application to tracking problems in celestical mechanics dates back to I. Newton [A.3].

The higher order terms of the series expansion of the solution (A.21) may be derived by inserting the ansatz

$$x(k) = (\gamma_1 + x_1(k)).k^{\beta_1} \quad , \quad x_1(o) = o$$

or

$$x(k) = \gamma_1.k^{\beta_1} + x_1(k).k^{\beta_1} \tag{A.24a}$$

into the relation (A.9) and by computing the small solutions $x_1(k) \in K^*\{k\}$ from the relation $F_1(x_1,k) = o$, where

$$F_1(x_1,k) := F(k,(\gamma_1 + x_1(k)).k^{\beta_1}) = o. \tag{A.24b}$$

The validity of this substitution will be discussed in the following lemma.

LEMMA A.4 :

Let $\beta_1 := \tan \alpha_1$ be the slope of a straight line $\mathcal{L}_1$ of the Newton Polygon of (A.9). Let γ_1 be an s_1-fold root of the corresponding supporting polynomial $\psi_j(\gamma)$ and let

$$h_1 := \min(\rho_i + i.\beta_1)$$

be the height of $\mathcal{L}_1$ (compare Fig. A.1) .

Then the formal power series (A.24b) has exactly $s_1 := s(F_1)$ small solutions $x_1(k)$ and h_1 trivial branching solutions starting at the point $(x,k) = (o,o)$, where $h_1 = h(F_1)$ and $s_1 := s(F_1)$ (compare Fig. A.2) . ${}_1\mathcal{L}$ is the Newton polygon of $F_1(x_1(k),k) = o$.

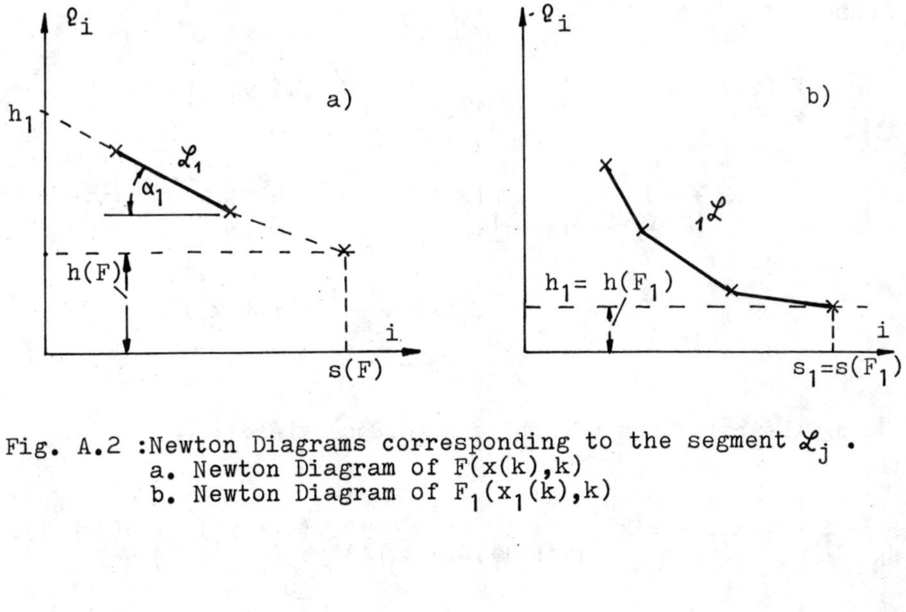

Fig. A.2 :Newton Diagrams corresponding to the segment $\mathcal{L}_j$.
a. Newton Diagram of $F(x(k),k)$
b. Newton Diagram of $F_1(x_1(k),k)$

Proof :

On the given assumption (γ_1 is an s_1-fold root of $\psi_j(\gamma)$) the supporting polynomial $\psi_j(\gamma)$ may be represented as

$$\psi_j(\gamma) = (\gamma - \gamma_1)^{s_1} . \tilde{\psi}_j(\gamma) \quad , \quad \tilde{\psi}_j(\gamma_1) \neq o \quad . \qquad (A.25)$$

Inserting the relation

$$x(k) = (\gamma_1 + x_1(k)).k^{\beta_1}$$

into (A.9) yields the expressions

$$F(x(k),k) = \sum_{i=0} F_i(k).k^{i.\beta_1}.(\gamma_1 + x_1)^i =: F_1(x_1,k) \quad (A.26)$$

or

$$F_1(x(k),k) = \sum_{\{i:\varrho_i+i.\beta_1=h_1\}} F_i(k).k^{i.\beta_1}.(\gamma_1 + x_1)^i$$

$$+ \sum_{\{i:\varrho_i+i.\beta_1>h_1\}} F_i(k).k^{i.\beta_1}.(\gamma_1 + x_1)^i$$

and

$$F_1(x_1(k),k) = k^{h_1}. \sum_{\{i:\varrho_i+i.\beta_1=h_1\}} F_{i,o}.(\gamma_1 + x_1)^i$$

$$+ \sum_{\{i:\varrho_i+i.\beta_1=h_1\}} (F_i(k) - F_{i,o}.k^{\varrho_i}).k^{i.\beta_1}.(\gamma_1 + x_1)^i$$

$$+ \sum_{\{i:\varrho_i+i.\beta_1>h_1\}} F_i(k).k^{i.\beta_1}.(\gamma_1 + x_1)^i \quad .$$

Using the relations (A.22b) and (A.23c) yields

$$F_1(x_1,k) = k^{h_1}. \sum_{\{i:\varrho_i+i.\beta_1=h_1\}} F_{i,o}.(\gamma_1 + x_1)^i + o(|k|^{h_1}).$$

Taking into account (A.22c) yields the relation

$$k^{-h_1}.F_1(x_1,k) = \psi_j(\gamma_1 + x_1) + o(|k|^{h_1}),$$

where h_1 is the height of the segment $\mathcal{L}_1$ of the corresponding Newton Polynom. Inserting the relation $\gamma = \gamma_1 + x_1$ into (A.25) yields the final expressions

$$\psi_j(\gamma_1+x_1) = x_1^{s_1}\cdot\tilde{\psi}_j(\gamma_1+x_1) \quad , \quad \tilde{\psi}_j(\gamma_1) \neq 0$$

and

$$F_1(x_1,k) = k^{h_1}.x_1^{s_1}.\{\tilde{\psi}_j(\gamma_1+x_1) + o(|k|^{h_1})\} \quad ,$$

where

$$\tilde{\psi}_j(\gamma_1+x_1)\Big|_{x_1=0} = \tilde{\psi}_j(\gamma_1+x_1(k))\Big|_{k=0} \neq 0 \,.$$

Therefore a term $k^{h_1}.x^{s_1}$ of smallest exponents h_1 and s_1 may be factored out from $F(x_1,k)$. These exponents h_1 and s_1 are the height $h(F_1)$ and the degree $s(F_1)$ of the power series $F_1(x_1,k)$.

LEMMA A.5 :

All small solutions $x(k)$ of (A.9) may be written in the form

$$x(k) = \sum_{j=1}^{\infty} \gamma_j.k^{\beta_j} \text{ , where } \beta_{i_1} < \beta_{i_2} \text{ for } i_1 < i_2 \,. \tag{A.27a}$$

Proof :

The small solutions of $F_1(x_1,k) = 0$ may again be computed by means of Lemma A.3. In case of a positive degree of degeneration $l_0({}_1\mathscr{L})$ of ${}_1\mathscr{L}$, equation (A.24b) has for $s_1 > 1$ at least one trivial solution $x_1(k) \equiv 0$ which implies a small solution of (A.9) of the form

$$x(k) = \gamma_1.\, k^{\beta_1} \text{ (finite power series) .}$$

For $l({}_1\mathscr{L}) > 0$, equation (A.24b) has exactly $l({}_1\mathscr{L})$ non trivial solutions of the form

$$x_{1j}(k) = \gamma_{2j} \cdot k^{\overline{\beta}_2} + o(|k|^{\overline{\beta}_2}) , \quad j = 1, \ldots, l({}_1\mathcal{L}) ,$$

where $\overline{\beta}_2$ is one of the slopes of the segment ${}_1\mathcal{L}$ and the numbers γ_{2j} are non vanishing (real or complex conjugate) roots of the corresponding supporting polynomial.
Introducing the abbreviations

$$\beta_2 := \beta_1 + \overline{\beta}_2 \quad \text{resp.} \quad \overline{\beta}_2 = \beta_2 - \beta_1 , \text{ where } \beta_1 < \beta_2 , \tag{A.27b}$$

the corresponding small solutions of (A.9) take the form

$$x(k) = \gamma_1 \cdot k^{\beta_1} + \gamma_2 \cdot k^{\beta_2} + o(|k|^{\beta_2}) .$$

After having determined the exponents β_1 to β_k and the coefficients γ_1 to γ_k, the exponent $\overline{\beta}_k$ has to be computed from the power series

$$F_k(x_k, k) = F_{k-1}((\gamma_k + x_k) \cdot k^{\beta_k}, k) , \tag{A.27c}$$

where $\overline{\beta}_k := \beta_k - \beta_{k-1}$ (compare (A.27b) .
For $l_o({}_k\mathcal{L}) > o$, the <u>finite</u> series

$$x(k) = \gamma_1 \cdot k^{\beta_1} + \ldots + \gamma_k \cdot k^{\beta_k}$$

is an exact small solution of (A.9). For $l({}_k\mathcal{L}) > o$, there exists at least one small solution of the form

$$x(k) = \sum_{i=1}^{k+1} \gamma_i \cdot k^{\beta_i} + o(|k|^{\beta_{k+1}}) ,$$

where $\overline{\beta}_{k+1} := \beta_{k+1} + \beta_k > \beta_k$, β_{k+1} is one of the slopes of the segment ${}_k\mathcal{L}$, and γ_{k+1} is a non vanishing root of the corresponding supporting polynomial.

The whole procedure may be continued to infinity.

The exponents $\beta_1, \ldots, \beta_j, \ldots$ as well as the coefficients $\gamma_1, \ldots, \gamma_j, \ldots$ are computed successively from a sequence of formal power series of the form

$$F(x,k)\ ,\ F_1(x,k)\ ,\ \dots\ ,\ F_j(x,k)\ ,\ \dots\ .$$

This sequence has to be computed for each of the small solutions (A.27a) .

Obviously, the set of all small solutions (A.27a) of (A.9) has exactly $s(F)$ elements some of which are equal. The question concerning the number of equal solutions of this set may be answered by means of the concept of multiplicity of small solutions of (A.9).

DEFINITION A.2 :

a. Assume there exists a small solution of (A.9) which may be represented in terms of a finite series of the form

$$x_j(k) = \sum_{i=1}^{m} \gamma_i \cdot k^{\beta_i} \quad , \quad m < \infty \ . \tag{A.28a}$$

Then there exist $l_o({}_m\mathscr{L})$ equal solutions of (A.9). The number $l_o({}_m\mathscr{L})$ is called multiplicity $v_j(x)$ of these solutions.

b. Assume there exists a small solution of (A.9) of the form

$$x_j(k) = \sum_{i=1}^{\infty} \gamma_i \cdot k^{\beta_i} \quad (\text{infinite series}) \ . \tag{A.28b}$$

Then we have

$$s(F) \geq s_1 = s(F_1) \geq \dots \geq s_k = s(F_k) \geq \dots > 0$$

for arbitrary values of the index k and for all power series $F_j(x_j,k)$. The multiplicity $v_j(x)$ of this solution is defined as

$$v_j(x) := \min_{i \in \mathbb{N}} (s_i) \quad .$$

In this case, there exists a finite number $i \in \mathbb{N}$ such that all supporting polynomials corresponding to a power series $F_l(x_l,k)$, $l \geq i$, have an $v_j(x)$-fold non vanishing root γ_l.

The following corollary is a direct consequence of the above definition.

COROLLARY A.6 :

Let $s(F) > 0$. Then we have

$$s(F) = \sum_{j=0}^{p} v_j(x) \quad , \tag{A.29a}$$

where p is the number of different small solutions of (A.9).

Proof :

According to lemma A.3 there exist $l_0(\mathscr{L})$ trivial solutions and $l(\mathscr{L})$ non trivial small solutions of (A.9). Then from definition A.1 we have , using the relation $s(F)= l_0(\mathscr{L})+l(\mathscr{L})$,

$$s(F) = l_0(\mathscr{L}) + l(\mathscr{L}) = \sum_{j=0}^{p} v_j(x) \quad ,$$

where $v_0(x) = l_0(\mathscr{L})$.
Let there exist r different segments $\mathscr{L}_j$ of the Newton Polygon corresponding to (A.9). Then the degree of the supporting polynomial of $\mathscr{L}_j$ is equal to $l(\mathscr{L}_j)$. For each segment $\mathscr{L}_j$ there exists exactly a number β_j and there exist exactly $l(\mathscr{L}_j)$ coefficients γ_{ji} and $l(\mathscr{L}_j)$ small nontrivial solutions (equal, different or partly different) of the form (A.28a) and (A.28b).The total number of small non trivial solutions of (A.9) is therefore equal to

$$l(\mathscr{L}) = \sum_{j=1}^{r} l(\mathscr{L}_j) \quad . \tag{A.29b}$$

If there exists a non vanishing zero γ_j of multiplicity $s_{1j} > 1$ of a corresponding supporting polynomial $\psi_j(\gamma)$, the corresponding term of s_{1j} of the series of small solutions is equal, and the multiplicity of these solutions has to be determined by means of the higher order terms of the corresponding series. In general we have r_1 different small solutions of multiplicity v_{j1} to v_{jr_1} , where $r_1 = r$ and where

$$s_{1j} = \sum_{i=1}^{r_1} v_{ji} \quad ,\text{and} \quad v_{ji} \le s_{1j} \text{ for } i=1, \ldots, r_1 .$$

Following these lines, collecting all different small solutions according to their multiplicities and using the relation

$$s = \sum_{j=1}^{r_1} s_j + l_o(\mathcal{L})$$

yields the relation (A.28a).

LEMMA A.7 :

Each formal power series of the form

$$x(k) = \sum_{j=1}^{\infty} \gamma_j . k^{\beta_j} , \quad \text{where} \quad 0 < \beta_1 < \beta_2 < \ldots \qquad \text{(A.27a)}$$

, and where (A.27a) is a small solution of (A.9) is an element of $K^{\infty}\{k\} := \bigsqcup_r K^r\{k\}$.

Proof :

In case of a finite series (A.27a), the statement of this lemma is already proved. All finite series of this form correspond to $K^{\infty}\{k\}$. Let x(k) be an infinite series of the form (A.27a) with multiplicity v. Let j_o be a finite number such that for all $j \ge j_o$ we have : $F_j(x_j, k) = 0$.
Then the corresponding Newton Polygon $_j\mathcal{L}$ contains exactly one segment $_j\mathcal{L}_1$. It has the lenght $l(_j\mathcal{L}) =: v$, and the supporting polynomial corresponding to $x_j(k)$ and to $F_j(x_j, k)$ has the form

$$\psi_j(\gamma) = \gamma^{\beta_j} . (\gamma - \gamma_j)^v \quad , \quad \gamma_j \ne 0 \quad . \qquad \text{(A.30a)}$$

According to lemma A.2 , this polynomial may be represented as

$$\psi_j(\gamma) = \gamma^{\beta_j} . \phi(\gamma^{q_j}) \quad ; \quad \text{degree}(\phi) = l(_j\mathcal{L})/q_j \quad , \qquad \text{(A.30b)}$$

where the numbers q_j are successively determined by the relation

$$ß_j - ß_{j-1} = \frac{p_j}{r.q_1. \dots .q_j} \qquad (A.30c)$$

with q_j and p_j relative prime.
The last relation follows by successively applying (A.27b). In connection with (A.27b) we have

$$ß_j = \bar{ß}_j + ß_{j-1} = \bar{ß}_j + \bar{ß}_{j-1} + ß_{j-2} = \dots = \bar{ß}_j + \dots + \bar{ß}_2 + ß_1$$

By using the relation $ß_1 = \bar{ß}_1$ we have

$$ß_j = \sum_{v=1}^{j} \bar{ß}_v \quad ; \quad ß_v , \bar{ß}_v > 0 \qquad (A.30d)$$

which yields the relation

$$ß_{i_1} > ß_{i_2} \quad \text{for} \quad i_1 > i_2 \quad .$$

From (A.13) we have

$$ß_1 = \bar{ß}_1 = p_1/r.q_1 \qquad \text{or}$$

$$ß_2 = ß_1 + \bar{ß}_2 \quad , \quad \bar{ß}_2 := l_2/q_2 \quad \text{and}$$

$$ß_2 = p_1/(r.q_1) + l_2/q_2 = (p_1 + \frac{r.q_1.l_2}{q_2})/(r.q_1) \quad .$$

In analogy to the following relations

$$p_2 := r.q_1.l_2 \quad ,$$

$$ß_2 = (p_1 + p_2/q_2)/(r.q_1) \quad ,$$

$$ß_3 = ß_1 + \bar{ß}_2 + \bar{ß}_3 \quad , \text{ where } \quad \bar{ß}_3 := l_3/q_3 \quad \text{or}$$

$$ß_3 = \frac{1}{r.q_1} .(p_1 + \frac{1}{q_2} .(p_2 + \frac{r.q_1.q_2.l_3}{q_3})) \quad ,$$

$p_3 := r.q_1.q_2.l_3$, where

$$\text{ß}_3 = \frac{1}{r \cdot q_1} \cdot (p_1 + \frac{1}{q_2} \cdot (p_2 + \frac{p_3}{q_3})) \quad ,$$

$$\text{ß}_4 = \frac{1}{r \cdot q_1} \cdot (p_1 + \frac{1}{q_2} \cdot (p_2 + \frac{1}{q_3} \cdot (p_3 + \frac{p_4}{q_4}))) \quad \text{, where}$$

$$p_4 := r \cdot q_1 \cdot q_2 \cdot q_3 \cdot l_4 \quad , \quad \overline{\text{ß}}_4 := \frac{l_4}{q_4} \quad ,$$

we have

$$\begin{aligned} \text{ß}_j &= \frac{1}{r \cdot q_1} \cdot (p_1 + \frac{1}{q_2} \cdot (p_2 + \dots + \frac{1}{q_{j-1}} \cdot (p_{j-1} + \frac{p_j}{q_j}) \dots) \\ &= \frac{p_j + p_{j-1} \cdot q_j + p_{j-2} \cdot q_{j-1} \cdot q_j + \dots + p_1 \cdot \prod_{\nu=2}^{j} q_\nu}{r \cdot \prod_{\nu=1}^{j} q_\nu} \end{aligned}$$

and

$$\begin{aligned} \overline{\text{ß}}_j &= \text{ß}_j - \text{ß}_{j-1} \\ &= \frac{p_j + p_{j-1} \cdot q_j + p_{j-2} \cdot q_{j-1} \cdot q_j + \dots + p_1 \cdot \prod_{\nu=2}^{j} q_\nu}{r \cdot \prod_{\nu=1}^{j} q_\nu} \\ &- \frac{(p_{j-1} + p_{j-2} \cdot q_{j-1} + \dots + p_1 \cdot \prod_{\nu=2}^{j-1} q_\nu) \cdot q_j}{r \cdot \prod_{\nu=1}^{j} q_\nu) \cdot q_j} \end{aligned}$$

or

$$\overline{\text{ß}}_j = \frac{p_j}{r \cdot \prod_{\nu=1}^{j} q_\nu} = \frac{l_j \cdot r \cdot \prod_{\nu=1}^{j-1} q_\nu}{r \cdot \prod_{\nu=1}^{j} q_\nu} = \frac{l_j}{q_j} \quad .$$

Comparing the relations (A.3oa) and (A.3ob) we have

$$v = \text{degree}\,\phi \quad , \quad v = l(\,_j\mathcal{L}) \quad \text{and}$$

$$\phi(\gamma^{q_j}) = (\gamma - \gamma_j)^{\,v} \quad ,$$

which yields the relation

$$q_j = 1 \quad \text{for all} \quad j \geq j_o \quad .$$

Therefore all variables $ß_j$ have the same denominator

$$r^* := r.\prod_{\nu=1}^{j_o} q_\nu \quad \text{for} \quad j \geq j_o \ ,$$

(compare (A.3oc)), which yields

$$ß_j = \frac{t_j}{r^*} \quad , \quad t_j \in \mathbb{N} \ .$$

Therefore all solutions of the form (A.21) are elements of $K^{r^*}\{k\}$.

It remains to be shown that all series expansions of the form (A.21) are solutions of (A.9).

The realtion (A.27c) yields

$$o = F(x(k),k) = F_1(x_1(k),k) = \ldots = F_j(x_j(k),k) \ ,$$

where

$$x_j(k) = k^{-ß_j}.(x(k) - \gamma_1.k^{ß_1} - \ldots - \gamma_j.k^{ß_j}) \quad .$$

Then for $j \in \mathbb{N}$ we have

$$F(x(k),k) = o(|k|^{h_j}) \quad ,$$

where h_j is the height of $F_j(x_j(k),k)$, and (A.27a) is a small solution of (A.9) if the relation

$$\lim_{j\to\infty} h_j = \infty$$

holds.Otherwise there exists a number $j \geq j_o$ such that the series $F_j(x_j(k),k)$ does not converge for small real values of k (There exists a trivial bifurcation at the point $(x,k) = (o,o)$).

In connection with lemma A.2 , the relation

$$h_i - h(F) \geq ß_i \quad , \ i \geq 1 \qquad \text{(A.13)}$$

implies :

$$h_1 - h(F) \geq \beta_1 \quad ,$$

$$h_2 - h_1 \geq \bar{\beta}_2 = \beta_2 - \beta_1 \quad ,$$

$$\vdots \qquad \vdots$$

$$h_j - h_{j-1} \geq \bar{\beta}_j = \beta_j - \beta_{j-1} \quad ,$$

or using the relation $h_j \geq 0$ for $j \in \mathbb{N}$ yields

$$h_1 \geq h + \beta_1 \quad , \quad h > 0 \quad ,$$

$$h_2 \geq h_1 + \beta_2 - \beta_1 \geq h + \beta_2 \geq \beta_2$$

$$h_3 \geq h_2 + \beta_3 - \beta_2 = h + \beta_3 + \beta_2 - \beta_2 \geq \beta_3 \quad ,$$

$$\vdots \qquad \vdots$$

$$h_j \geq h + \beta_j \geq \beta_j \quad .$$

The relations $\beta_j = \sum_{\nu=1}^{\infty} \beta_\nu$ and $\beta_j = \beta_{j_0}$ for $j \geq j_0$ imply

$$\lim_{j\to\infty} \beta_j = \infty \quad , \quad \text{and using the relation} \quad h_j \geq \beta_j \ ,$$

$$\lim_{j\to\infty} h_j = \infty \quad .$$

From (A.27b) we have

$$0 < \beta_{i_1} < \beta_{i_2} \quad \text{for} \quad i_1 < i_2 \quad .$$

Using (A.30c) and (A.30d) there exist numbers $i, r \in \mathbb{N}$,such that each β_j may be represented as $\beta_j = i/r$.Therefore each small solution of (A.9) may be represented in the form of the relation (A.31)

$$x(k) = \sum_{j=1}^{\infty} \gamma_j \cdot k^{j/r} \quad . \qquad \rfloor \tag{A.31}$$

Till now s(F) small solutions of(A.9) have been constructed which belong to $K^{\infty}\{r\}$. It remains to be shown that the set of these small solutions may be grouped into subsets of conjugate solutions.

DEFINITION A.3 :

Let λ be a primitive r-fold root of the number 1 , i.e.

$$\lambda = (1)^{1/r} = (e^{2.\pi.i.n})^{1/r} = e^{2.\pi.i.n/r} ,$$

where i is the imaginary unit and $n \in \mathbb{N}$. Let x(k) be a solution of (A.9) of the form

$$x(k) = \sum_{j=0}^{\infty} c_j.k^{j/r} \quad . \tag{A.32a}$$

Then the series expansions

$$\begin{aligned} x_1(k) &= \sum_{j=1}^{\infty} c_j.\ \lambda^1.k^{j/r} \\ &\vdots \\ x_{r-1}(k) &= \sum_{j=1}^{\infty} c_j.\ \lambda^{(r-1)}.k^{j/r} \end{aligned} \tag{A.32b}$$

are called <u>conjugate</u> to (A.32a) .

LEMMA A.8 :

Let (a.31) be a solution of (A.9). Then all series expansions which are conjugate to (A.31) are also solutions of (A.9).

Proof :

Inserting (A.31) into (A.9) yields the relations

$$0 = F(x(k),k) = F(\sum_{j=1}^{\infty} \gamma_j.k^{j/r},k) = F(\sum_{j=1}^{\infty} \gamma_j.(1.k)^{j/r},k)$$

and

$$o = F\left(\sum_{j=1}^{\infty} \gamma_j \cdot (1)^{j/r} . k^{j/r}, k\right) = F\left(\sum_{j=1}^{\infty} \gamma_j \cdot \lambda^j . k^{j/r}, k\right)$$

$$= F\left(\sum_{j=1}^{\infty} \gamma_j (1.1)^{j/r} . k^{j/r}, k\right) = F\left(\sum_{j=1}^{\infty} \gamma_j \cdot \lambda^{2.j} . k^{j/r}, k\right)$$

$$= \ \ldots = F\left(\sum_{j=1}^{\infty} \gamma_j \cdot (1. \ \ldots \ .1) . k^{j/r}, k\right)$$

$$= F\left(\sum_{j=1}^{\infty} \gamma_j \cdot \lambda^{\mu . j} . k^{j/r}, k\right) \ , \ \mu = 1 \ , \ \ldots \ , \ r-1 \ ,$$

where

$$\lambda = (1)^{1/r} \quad \text{or} \quad \lambda^{\mu \cdot j} = (1)^{(\mu . j)/r} \quad \text{or}$$

$$\lambda = (e^{2.\pi.i.n})^{1/r} = e^{(2.\pi.i.n)/r} \ , \ i = 1, 2, \ \ldots \ .$$

The set of conjugate solutions of (A.9) yields a group of small solutions of (A.9). Such a group of small solutions of (A.9) will for instance be generated by a segment of the Newton Polygon of length bigger than 1 meeting only two points of the Newton Diagram. ┘

The following corollary is an obvious consequence of the foregoing considerations. It will be stated without proof.

COROLLARY A.9 :

a. A small solution (A.27a) of (A.9) has the multiplicity one iff the following relation holds:

$$\partial F(x(k),k)/\partial x \neq o \quad \text{for} \quad (x,k) = (o,o).$$

b. The multiplicity of (A.27a) is equal to the smallest order j of the non vanishing terms

$$\partial F(x(k),k)/\partial x \ , \ \ldots \ , \ \partial^j F(x(k),k)/\partial x^j ; (x,k) = (o,o).$$

c. Let s:= s(F) be the degree of F(x,k) (comp. definition A.1). Then the relation (A.9) is locally near the point (x,k)=(o,o) algebraic equivalent to an equation of the form

$$x^s + a_1(k).x^{s-1} + \dots + a_s(k) = o \quad , \qquad (A.33)$$

where $a_i(k) \in K^*\{k\}$ for $i=1,2, \dots ,s$.

The relations (A.9) and (A.33) have identical small solutions $x(k)$ of the same degree of multiplicity.

The practical importance of the Newton Diagram Technique is based upon the fact that the first terms of the series expansion of the small solutions of (A.9) may be computed graphically from the knowledge of only a finite set of coefficients of (A.9).

A.2 Analytic Functions

Theorem A.1 in connection with Lemma A.2 to corollary A.9 provide rules for constructing all small roots of the formal power series of (A.9). Within this general framework, convergence statements of the small solutions of (A.9) cannot be made. In order to get convergence statements of the small solutions of (A.9), we have to restrict ourselves to analytic functions F(x,k) or to pseudopolynomials of the form (A.34a) which are of some importance in connection with the computation of the higher order terms of the small solutions.

$$F(x,k') = \sum_{i=o}^{n} F_i(k').x^i = o \quad , \qquad (A.34a)$$

where

$$F_i(k') = k'^{\varrho_i}. \sum_{j=1}^{\infty} F_{ij}.k'^{j/r} \qquad \text{and}$$

$$x \in \mathbb{C}^1 \; ; \; k, F_{ij} \in \mathbb{R}^1 \quad \text{and} \quad \varrho_i \in \mathbb{Q}^+ \; .$$

Replacing the term $k^{1/r}$ by means of the term k transforms the relation (A.34a) into an analytic function of the form

$$F(x,k) = \sum_{i,j=1} F_{ij} . x^i . k^j = 0 \quad . \qquad (A.34b)$$

Again, the small solutions of (A.34b) may be constructed by means of the Newton Diagram corresponding to (A.34b). In agreement with Chapter A.1 they have the form

$$x(k) = \sum_{j=1} \gamma_j . k^{\beta_j} \quad . \qquad (A.35)$$

Then the following theorem holds:

THEOREM A.1o :

The small solutions of (A.34b) of the form (A.35) are convergent power series near the point (x,k) = (o,o) .

This theorem will be proved on the basis of the (one-dimensional) Weierstrass Preparation Theorem which may be restated as follows [A.4] :

THEOREM A.11 (Weiertsrass Preparation Theorem) :

Let $F : \mathbb{C}^1 \times \mathbb{R}^1 \longrightarrow \mathbb{C}^1$ be an analytic function

$(x,k) \longmapsto F(x,k)$

near the point (x,k) = (o,o) .

Let $\partial^i F / \partial x^i \Big|_{(x,k)=(0,0)} = 0 \quad , \quad i = 0,1, \ldots, r-1.$

and

$\partial^r F / \partial x^r \Big|_{(x,k)=(0,0)} \neq 0 \quad$ for $r > 0$.

Then the function F(x,k) may be represented near the point (x,k) = (o,o) in the form

$$F(x,k) = \{ x^r + h_1(k).x^{r-1} + \dots + h_{r-1}(k).x + h_r(k) \} .\Omega(x,k), \qquad (A.36)$$

where the functions $h_i(k)$, $i = 1, \dots, r$, are analytic in k near the point $(x,k) = (0,0)$ and satisfy the relations $h_i(0) = 0$. The function $\Omega(x,k)$ is analytic in x and k and satisfies the relation $\Omega(x,k) \neq 0$ near the point $(x,k)=(0,0)$.

The function $\Omega(x,k)$ behaves near the point $(x,k) = (0,0)$ as a non vanishing constant. Therefore, the local behaviour of $F(x,k)$ near the point $(x,k) = (0,0)$ may be described by means of a polynomial of degree r in x with coefficients which are analytic in k.

Proof of theorem A.1o :

According to the Weierstrass Preparation Theorem for scalar analytic functions, the relation (A.34b) is locally (for k=o) equivalent to an equation of the form

$$F(x,k) = G(x,k).Q(x,k) = 0 \quad , \qquad (A.37a)$$

where $G(x,k)$ may be written as

$$G(x,k) = x^s + H_{s-1}(k).x^{s-1} + \dots + H_1(k).x + H_0(k) \; , \qquad (A.37b)$$

where the functions $H_i(k)$ are analytic in k, $H_i(0)=0$ for $i = 0, \dots, s-1$, $Q(x,k)$ is analytic in x and k and $Q(0,0) \neq 0$. The small solutions of (A.37a) are equivalent to the small solutions of the relation

$$G(x,k) = 0 \quad . \qquad (A.37c)$$

The close connection of $G(x,k)$ to the supporting polynomial (A.12) is obvious.

According to a Theorem of Rouché [A.1], there exist positive numbers ε and δ such that equation (A.37c) has for $|k| < \delta$ exactly s small solutions

$$x_1(k) \; , \; \dots \; , \; x_s(k)$$

satisfying the inequality $|x_i(k)| < \varepsilon$ for $i = 1, \ldots, s$.
Despite the fact that the function $G(x,k)$ is uniquely determined by Theorem A.11, this information will not be needed here in full detail. Instead, the small solutions of (A.34b) will again be computed by means of the Newton Diagram Technique.

Starting from a segment $\mathcal{L}_j$ of the Newton Polygon with slope $\beta_1 = p_1/q_1$ and with endpoints $(\lambda, \varrho_\lambda)$ and $(\varkappa, \varrho_\varkappa)$, $(\lambda < \varkappa)$, and using the relations

$$\begin{aligned} \varkappa - \lambda &= \mu \cdot q_1 \quad, \varkappa > \lambda, \; \mu \in \mathbb{N}, \\ \varrho_\lambda - \varrho_\varkappa &= \mu \cdot p_1 \quad, \\ \beta_1 &= p_1/q_1 \end{aligned} \tag{A.38a}$$

and

$$x = y^{p_1} \cdot \gamma_1 \quad, \quad \text{where} \quad k = y^{q_1} \quad, \tag{A.38b}$$

the equation (A.34b) may be transformed locally to the form

$$F(x,k) = \sum_{i=0}^{\infty} k^{\varrho_i} \cdot x^i \cdot F_{i,o} + o(|k|^{\varrho_i} + |x|^i) \tag{A.38c}$$

$$= \sum_{i=0}^{\infty} F_{i,0} \cdot y^{q_1 \cdot \varrho_i} \cdot y^{p_1 \cdot i} \cdot \gamma_1^i + o(|k|^{\varrho_i + i \cdot \beta_1})$$

$$= \sum_{i=0}^{\infty} F_{i,0} \cdot \gamma_1^i \cdot y^{p_1 \cdot i + q_1 \cdot \varrho_i} + o(|y|^{q_1 \cdot \varrho_i + i \cdot p_1})$$

$$=: \tilde{\psi}(y,\gamma) = 0 \quad,$$

where $\tilde{\psi}(y,\gamma)$ is analytic with respect to y and to γ. For $y=0$ the function $\tilde{\psi}(y,\gamma)$ corresponds to the supporting polynomial $\psi(\gamma)$ of (A.12). Then two cases have to be treated:

Case 1:

The polynomial $\tilde{\psi}(y,\gamma)$ has only simple roots γ_{1i} for $y=0$. Then we have

$$\psi(y,\gamma_{1i})\Big|_{y=0} = 0 \quad \text{and} \quad \partial\psi(y,\gamma_{1i})/\partial\gamma\Big|_{y=0} \neq 0 \;. \qquad (A.38d)$$

According to the Implicit Function Theorem for analytic functions the equation (A.38c) has the convergent power series (A.38e) as a unique small solution.

$$\gamma_1(y) = \gamma_{1i} + \sum_{j=1}^{\infty} c_{ij}.y^j \qquad (A.38e)$$

$$x = \gamma_{1i}(y).y^{p_1} = \gamma_{1i}.k^{p_1/q_1} + \sum_{j=1}^{\infty} c_{ij}.k^{j/q_1}.k^{p_1} ,$$

$$x(k) = \gamma_{1i}.k^{\beta_1} + \sum_{j=1}^{\infty} c_{1j}.k^{\beta_1 . j} \;.$$

Case 2 :

The polynomial $\widetilde{\psi}(y,\gamma)$ has a solution $x(k)$ of multiplicity s_1. Then according to chapter A.1 there exists a positive number $s < s_1$ such that the corresponding supporting polynomial $\psi_i(\gamma\ )$ has an s-fold root γ_i for $i \geq i_o$, and the equation (A.34b) has an s-fold small solution

$$x_1(k) = \sum_{i=1}^{i_o} \gamma_i.k^{\beta_i} + \sum_{i=1_o+1} \gamma_i.k^{\beta_i} \;. \qquad (A.38f)$$

Assume $s > 1$. Then the equation (A.37c) may be written in the form

$$G(x,k) = g_1(x,k)^s.G_1(x,k) \quad ,$$

where $G_1(x_1(k),k) \neq 0$ near the point $(x,k) = (0,0)$ and $x_1(k)$ is a simple small solution of the irreducible polynomial $g_1(x,k) = 0$. Following the same reasoning as in case 1 proves the convergence of a small multiple solution of (A.34b). ┘

The convergency questions of the small solutions of the pseudopolynomial (A.34a) may be treated in analogy to the corresponding questions in connection with (A.34b) by replacing the neighbourhood of the point $(x,k) = (o,o)$ by the concept of a punctuated neighbourhood of this point [A.5].

If instead of analytic functions the small solutions of C^{∞}-functions are to be investigated, the Weierstrass Preparation Theorem as a basic tool of analytic function theory has to be replaced by its equivalent, the <u>Malgrange Preparation Theorem</u> [A.6] . This later theorem serves as a basis for a classification of C^{∞}-functions and is the formal idea of the <u>Catastrophe Theory</u> of R. Thom [A.7] .The first steps of a procedure for constructing the small solutions of the vector case corresponding to equation (A.9) are treated in [A.8] .

References

[A.1] E. Hille , Analytic Function Theory , Vol. II , Ginn , 1972 .

[A.2] M. A. Krasnoselskii et al. , Approximate Solutions of Operator Equations , Noordhoff, 1972 .

[A.3] D.T. Whiteside , The Mathematical Papers of Isaac Newton, Vol.III, Cambridge University Press,1969.

[A.4] S. Bochner , W.T. Martin, Several Complex Variables, Princeton University Press , 1948.

[A.5] H. Hahn, Zur Theorie und Technik singulärer Regelkreise, Habilitationsschrift, Universität Tübingen, Fachbereich Physik, 1977/78, Chapter III.

[A.6] G. Wassermann, Stability of Unfoldings, Springer Lect. Not. in Math. , No. 393, 1975.

[A.7] R. Thom , Stabilité Structurelle et Morphogenèse,
Benjamin Inc., 1972 .

[A.8] H. Hahn , compare [A.5] , Appendix II .

Index

Spectral Characterization of Controllability of Linear Time-invariant Systems under Control Restraints

Bernhard Herz

Summary

For linear time-invariant systems under control restraints, $\dot{y}(t) = A\,y(t) + B\,v(t)$, $t \in \mathbb{R}$, $y \in \mathbb{R}^n$, $v \in \Omega$ in the continuous-time case and
$x(k+1) = F\,x(k) + G\,u(k)$, $k \in \mathbb{N}$, $x \in \mathbb{R}^n$, $u \in \Omega$
in the discrete-time case, where $\Omega \subset \mathbb{R}^m$ is a bounded set which contains the origin in its interior, the problem of complete null-controllability is investigated. It is proved that such a system is completely null-controllable under the control restraints $v \in \Omega$ ($u \in \Omega$) if and only if the following two conditions hold: i. the system is completely null-controllable in the absence of control-restraints, that is $v \in \mathbb{R}^m$($u \in \mathbb{R}^m$) is admitted, ii. all eigenvalues of A have real parts less or equal to zero (all eigenvalues of F are less or equal to one in absolute value). This verifies a well-known assertion, stated without proof in 1960 by Kalman [1], extends a theorem of Pontryagin et al. [3] from 1961 and generalizes some special results, obtained in the meantime. A short review of the history of the problem is given in the introduction.

Zusammenfassung

Für lineare zeitinvariante Systeme mit Stellgrößenbeschränkung, $\dot{y}(t) = A\,y(t) + B\,v(t)$, $t \in \mathbb{R}$, $y \in \mathbb{R}^n$, $v \in \Omega$ bzw. $x(k+1) = F\,x(k) + G\,u(k)$, $k \in \mathbb{N}$, $x \in \mathbb{R}^n$, $u \in \Omega$, worin

$\Omega \subset \mathbb{R}^n$ beschränkt ist und den Ursprung im Innern enthält, wird die Frage der vollständigen Null-Steuerbarkeit untersucht. Es wird gezeigt, daß vollständige Null-Steuerbarkeit unter der Stellgrößenbeschränkung $v \in \Omega$ ($u \in \Omega$) genau dann vorliegt, wenn die folgenden beiden Bedingungen erfüllt sind: i. Das System ohne Stellgrößenbeschränkung ist vollständig nullsteuerbar. ii. Kein Realteil eines Eigenwertes von A ist echt positiv. (Die Beträge aller Eigenwerte von F sind kleiner oder gleich 1). Damit wird der Beweis zu einer schon 1960 durch Kalman [1] aufgestellten Behauptung nachgetragen, ein Satz von Pontryagin et al. [3] von 1961 erweitert und eine Reihe in der Zwischenzeit erzielter speziellerer Ergebnisse verallgemeinert. In der Einleitung wird ein kurzer Abriß der Geschichte des untersuchten Problems gegeben.

1. Introduction and some Remarks on the History of the Problem

The nowadays common idea of complete null-controllability of linear time-invariant systems $\dot{y}(t) = A\,y(t) + B\,v(t)$, (1c), in the continuous-time case, and $x(k+1) = F\,x(k) + G\,u(k)$, (1d), in the discrete-time case - for a more detailed specification see the beginning of chapter 2 - does not take into account any control restraints. But from the point of view of the applications it appears more natural to include some control restraints into consideration since in reality no control mechanism applied is able to generate unlimited controls. Therefore the problem of complete null-controllability of linear time-invariant systems under control restraints is being investigated almost since the introduction of the idea of usual controllability. As to the control restraints it is realistic to assume that the admissible controls v or u, respectively, are to be taken from a bounded set $\Omega \in \mathbb{R}^n$ containing the origin in its interior. All papers referred to below in a short review of the history of the problem make such an assumption in some form. In addition, througout the literature quoted, it is of course assumed - as it will be done in this paper as well - that the considered linear system is completely null-controllable in the absence of control restraints, since evidently the introduction of control restraints never enlarges the set of null-controllable states.

As the theorems 1c and 1d of Chapter 2 will show, complete null-controllability under the above assumption depends only on spectral properties of A or F, respectively. More precisely, as necessary and sufficient it turns out that all eigenvalues of A have real parts less or equal to zero or that all eigenvalues of F are less or equal to one in absolute value, respectively. This means, that every initial state of a linear system can be steered to the origin by means of a control being admissible in the sense above if and only if the linear system is not able to free motions of exponential growth. It should be pointed out that this condition does not exclude systems which are able to free motions of polynomial growth, as it occurs if A has eigenvalues on the imaginary axis or if F has eigenvalues on the unit circle line, respectively, with non-coinciding geometric and algebraic multiplicities in each case. Thus the theorems 1c and 1d do not formulate an up to now unknown statement, but supplement the proof of a well-known assertion, stated first in 1960 by Kalman [1], which generally is accepted and applied in the engineering science, as for instance by Ludyk [9].

Subsequently it is given a survey on the stepwise progress obtained since 1960 in proving Kalman's assertion.

In 1961 Pontryagin et al. [3] investigated the continuous-time case. In their famous book on 'The Mathematical Theory of Optimal Processes', Theorem 14, p. 130, they proved the sufficiency of the condition Re $\lambda_i(A) < 0$, where

$\lambda_i(A)$, $i = 1,\dots,n$ denotes the eigenvalues of A. Although some supplementary remarks on page 139 of the book show the author's further interest in the case of a nonstable A, the limit case (in the sense of Kalman's assertion) Re $\lambda_i(A) \le 0$ for $i = 1,2,\dots,n$ but Re $\lambda_i(A) = 0$ for some $i \in \{1,2,\dots,n\}$ is not treated. The obvious interest of the engineering science in this limit case is for instance illustrated by Theorem 6 - 10 in Athans and Falb [5] , page 420. In this theorem, classified by the authors on page 427 as a 'most useful existence theorem' they assert correctly the sufficiency of the condition Re $\lambda_i(A) \le 0$, but for a proof they refer to the above mentioned result of Pontryagin et al. [3].

The special case, that in the discrete-time system (1d) F is regular, u is a scalar, thus $G = g \in \mathbb{R}^n$, and $\Omega = \{u \mid |u| \le 1\} \subset \mathbb{R}^1$, was treated by Desoer and Wing ([4], Theorem 2) in 1961. But their proof of the sufficiency of the condition $|\lambda_i(F)| \le 1$ for $i = 1,\dots,n$ is not correct. As a counterexample to their proof as well as to their construction rule for an admissible control given within the proof the following system may be considered.

$$\begin{bmatrix} x_1 \\ x_2 \\ x_3 \end{bmatrix}(k+1) = \begin{bmatrix} 1 & -1 & 1 \\ 0 & 1 & -1 \\ 0 & 0 & 1 \end{bmatrix} \begin{bmatrix} x_1 \\ x_2 \\ x_3 \end{bmatrix}(k) + \begin{bmatrix} 0 \\ 0 \\ 1 \end{bmatrix} u(k)$$

For initial values $x^0 := \begin{bmatrix} 0 \\ \rho/2 \\ \rho \end{bmatrix}$ with $\rho > 0$,

sufficiently large, the proof does not apply and the construction rule does not yield an admissible control bringing the initial state to the origin within a finite number of steps. Nevertheless, the basic pattern of this proof -

for the first time applied in [4] - is obviously suitable for the problem and will be used in a modified form in this paper as well. The underlying idea is to test the effectivity of a given construction rule for an admissible control by consideration of a divergence problem for a certain vector- or matrixvalued series.

In 1967 Lee and Markus [7] , Theorem 8, page 92, gave for single-input continuous-time systems (1c) the first correct proof of Kalman's assertion including the critical case that $\mathrm{Re}\ \lambda_i(A) = 0$ for some i. In a corollary on page 96 they extended the result to those multi-input systems which can be effectively replaced by single-input systems. Their sophisticated proof of the sufficiency of $\mathrm{Re}\ \lambda_i(A) \leq 0$ does not admit further generalization, whereas the proof of the necessity of $\mathrm{Re}\ \lambda_i(A) \leq 0$ is easily extendable to the general case. The proof of the necessity of the condition $|\lambda_i(F)| \leq 1$ in the discrete-time case, being given in this paper, has been suggested by the corresponding proof in [7] for the continuous-time case, in which the study of certain differential inequalities is substituted by the study of certain difference inequalities.

In this paper the discrete-time case will be treated first. The main tool for the proof of the sufficiency of $|\lambda_i(F)| \leq 1$ for $i = 1,\ldots,n$ will be a certain consequence of Vieta's theorem applied to n,n-matrices having only eigenvalues equal to one in absolute value. The sufficiency of the condition $\mathrm{Re}\ \lambda_i(A) \leq 0$ for $i = 1,\ldots,n$ in the continuous-time case then easily will be obtained from the discrete-time result by means of well-known theorems concerning the preservation of controllability under sampling.

2. Notation, Statement of Problem, Main Results

We consider linear time-invariant systems; a discrete-time one

$$x(k + 1) = F\,x(k) + G\,u(k) \quad ; \quad k \in \mathbb{N}_o \tag{1d}$$

and a continuous-time one

$$\dot{y}(t) = A\,y(t) + B\,v(t) \quad ; \quad t \in \mathbb{R} \tag{1c}$$

where $x,y \in \mathbb{R}^n$ are the state vectors; $u,v \in \mathbb{R}^m$ are the control vectors; F, G, A and B are real constant matrices of appropriate dimensions. The solution of (1d) with initial condition $x(k_o) = x^o$, $k_o \in \mathbb{N}_o$, under the control $u(k)$, $k = k_o, k_o + 1, \ldots$ will be denoted by $x(k;\, k_o, x^o,\, u(\cdot))$, $k = k_o$, $k_o + 1, \ldots$. Similary we will denote the solution of (1c) by $y(t;\, t_o,\, y^o,\, v\,(\cdot))$, $t \geq t_o$, where we always assume that the acting control $v(t)$ is piecewise continuous.

Since (1d) and (1c) are time-invariant, controllability properties may be formulated and investigated without loss of generality by setting $k_o = 0$ and $t_o = 0$.

Definition 1d

Let Ω be an arbitrary subset of $\mathbb{R}^m$. Then the system (1d) is called *null-controllable under the control restraint* $u \in \Omega$ if for any initial state x^o, there exists a finite integer $k_1 > 0$ and a control

$$u(\cdot) \; : \; \{0,1,\ldots,k_1 - 1\} \longrightarrow \Omega$$

such that $x(k_1;\, 0,\, x^o,\, u(\cdot)) = 0$

□

Definition 1c

For $\Omega \subset \mathbb{R}^m$ we call the system (1c) *null-controllable under the control restraint* $v \in \Omega$ if for any initial state y^o, there exists a finite $t_1 > 0$ and a piecewise continuous control

$$v(\cdot) : [0,t_1] \longrightarrow \Omega$$

such that $y(t_1; 0, y^o, v(\cdot)) = 0$

□

If in Definition 1d or 1c $\Omega = \mathbb{R}^m$, we will only speak of *null-controllability*.

Let us assume for convenience that $\mathbb{R}^m$ and $\mathbb{R}^n$ are fitted with the respective maximum-norms; that is for instance

$$\|u\| := \max_{i=1,\dots,m} |u_i| \quad \text{and} \quad \|x\| := \max_{i=1,\dots,n} |x_i|.$$

Consequently all matrix norms will be lub-norms with respect to these vector norms. For every $r > 0$ let $\Gamma_r^m := \{u \in \mathbb{R}^m \mid \|u\| \le r\}$. Hence a set $\Omega \subset \mathbb{R}^m$ is bounded and contains $u = 0$ in its interior if and only if there exist two values

$$0 < r_1 < r_2 < \infty, \text{ such that } \Gamma_{r_1}^m \subset \Omega \subset \Gamma_{r_2}^m. \quad (2)$$

The following two theorems deal with controllability properties when such sets $\Omega \subset \mathbb{R}^m$ figure as control restraints.

Theorem 1d

Let $\Omega \subset \mathbb{R}^m$ be bounded with $u = 0$ in its interior. Then the system 1d is null-controllable under the control restraint $u \in \Omega$ if and only if the following two conditions hold:

System 1d is null-controllable; (3d)

every eigenvalue λ of F satisfies $|\lambda| \le 1$ (4d)

□

Theorem 1c

Let $\Omega \subset \mathbb{R}^m$ be bounded with $v = 0$ in its interior.

Then the system (1c) is null-controllable under the control restraint $v \in \Omega$ if and only if the following two conditions hold:

System 1c is null-controllable; (3c)

every eigenvalue λ of A satisfies $\mathrm{Re}\,\lambda \leq 0$ (4c)

□

The proofs will be given in section 4.

3. Some Preliminaries

i. Evidently for the solutions of system (1d) the so called superposition-principle

$$\alpha x(k;\, k_o,\, x^o,\, u(\cdot)) + \beta x(k;\, k_o,\, \tilde{x}^o,\, u(\cdot)) =$$

$$x(k;\, k_o,\, \alpha x^o + \beta\tilde{x}^o,\, \alpha u(\cdot) + \beta\tilde{u}(\cdot))$$

is valid and for the solutions of system (1c) an analogous formula is true.

ii. Let $\Omega \subset \mathbb{R}^m$ have the property (2); and let $\delta > 0$ be arbitrary. Then the system (1d) is null-controllable under the control restraint $u \in \Omega$ if and only if it is null-controllable under the control restraint $u \in \Gamma^m_\delta$.

This is an immediate consequence of the linearity of system (1d): First we assume that system (1d) is null-controllable under the control restraint $u \in \Omega$. For an arbitrary but fixed state $x^o \in \mathbb{R}^n$ we define the auxiliary state $\hat{x}^o := r_2/\delta\, x^o$. There is an integer $k_1 > 0$ and a control $\hat{u}(k) \in \Omega \subset \Gamma^m_{r_2}$, $k = 0,1,\ldots,k_1 - 1$, such that

$x(k_1;\, 0,\, \hat{x}^o,\, \hat{u}(\cdot)) = 0$ and thus $x(k_1;\, 0,\, x^o,\, \delta/r_2\, \hat{u}(\cdot)) = 0$ But from $\hat{u}(k) \in \Gamma^m_{r_2}$ we have $\|\hat{u}(k)\| \leq r_2$ and therefore $u(k) := \delta/r_2\, \hat{u}(k) \in \Gamma^m_\delta$ for $k = 0, 1, \ldots, k_1 - 1$. Reciprocally, assuming the null-controllability of system (1d) under the control restraint $u \in \Gamma^m_\delta$, we associate to an arbitrary fixed $x^o \in \mathbb{R}^n$ the auxiliary state $\tilde{x}^o := \delta/r_1\, x^o$. Proceeding in the same manner as above we conclude that x^o can be steered to 0 under the control restraint $u \in \Gamma^m_{r_1} \subset \Omega$.

Obviously identical conclusions can be made in case of system (1c).

These simple considerations show that conditions for null-controllability under a control-restraint $u \in \Omega$, when Ω has property (2), do not depend on further specification of Ω. Consequently we may restrict ourselves for the purpose of proofs or even parts of proofs of the above Theorems 1d, 1c to the case $\Omega = \Gamma^m_\delta$ for some $\delta > 0$.

iii. Since the controllability properties under consideration are geometric or even physical properties, and since on the other hand the conditions (4d) and (4c) are invariant under linear regular transformations of the state-space $\mathbb{R}^n$, we may assume for the purpose of proofs that the matrices F in system (1d) and A in system (1c) are in real Jordan canonical form (see for example Kowalsky [8]).

iv. If F is singular we may assume that system (1d) is given in the following form:

$$\begin{bmatrix} x_1(k+1) \\ \cdots\cdots \\ x_2(k+1) \end{bmatrix} = \begin{bmatrix} F_1 & \vdots & 0 \\ \cdots & \vdots & \cdots \\ 0 & \vdots & F_2 \end{bmatrix} \begin{bmatrix} x_1(k) \\ \cdots\cdots \\ x_2(k) \end{bmatrix} + \begin{bmatrix} G_1 \\ \cdots \\ G_2 \end{bmatrix} u(k) \qquad (5)$$

where F_2 is a real n_2,n_2-matrix, $1 < n_2 \leq n$, with eigenvalues $\lambda_i = 0$ for $i = 1,\ldots,n_2$. Then we have $F_2^{n_2} = 0$. Obviously (5) can be considered as the two independent systems

$$x_1(k+1) = F_1\, x_1(k) + G_1\, u(k) \qquad (6)$$

with solution $x_1(k;\, k_o,\, x_1^o,\, u(\cdot))$, and

$$x_2(k+1) = F_2\, x_2(k) + G_2\, u(k) \qquad (7)$$

with solution $x_2(k;\, k_o,\, x_2^o,\, u(\cdot))$
steered by the same control. Thus the solution of (5), started at $k_o = 0$, can be written as

$$x(k;0, \begin{bmatrix} x_1^o \\ \cdots \\ x_2^o \end{bmatrix}, u(\cdot)) = \begin{bmatrix} x_1(k;0,x_1^o,u(\cdot)) \\ \cdots\cdots\cdots\cdots \\ x_2(k;0,x_2^o,u(\cdot)) \end{bmatrix}, \ k = 0,1,2,.. \qquad (8)$$

Let us now assume that for x_1^o there is a $k_1^* > 0$ and a control $u(k) \in \mathbb{R}$, $k = 0, 1, \ldots, k_1^* - 1$, such that $x_1(k_1^*;\, 0,\, x_1^o,\, u(\cdot)) = 0$. For system (5) this control yields

$$x(k_1^*;0, \begin{bmatrix} x_1^o \\ \cdots \\ x_2^o \end{bmatrix}, u(\cdot)) = \begin{bmatrix} 0 \\ \cdots \\ x_2^* \end{bmatrix} \quad \text{with some}$$

$x_2^* := x_2(k_1^*,\, 0\, ,\, x_2^o,\, u(\cdot)) \in \mathbb{R}^{n_2}$. Extending the so far applied control by $u(k) = 0$ for $k = k_1^*$, $k_1^* + 1, \ldots, k_1^* + n_2 - 1$ finishes the job,

$$x(k_1^* + n_2;\ 0, \begin{bmatrix} x_1^o \\ \dots \\ x_2^o \end{bmatrix}, u(\cdot)) = \begin{bmatrix} F_1^{n_2} & 0 \\ \dots & \dots \\ F_2^{n_2} & x_2^* \end{bmatrix} = \begin{bmatrix} 0 \\ \cdot \\ 0 \end{bmatrix} .$$

Thus the control problem for system (5) is reducible to a control problem for system (6). Furthermore it is evident by formula (8) that null-controllability of system (5) implies null-controllability of system (6). Thus if system (5) satisfies conditions (3d), (4d) the same is true for system (6). Combining all preceeding items it becomes clear that for the proof of the sufficiency part of theorem 1d we may restrict ourselves to the case $\det F \neq 0$.

v. The last preparatory remark deals with a simple consequence of Vieta's theorem to certain matrices.

For a natural number n let $\binom{n}{j}$, $j = 0, 1, \dots, n$, be the binomial coefficients and

$$\kappa(n) := 1/ \max_{j=0,1,\dots,n} \binom{n}{j}. \text{ Hence } \kappa(n) \in (0,1]. \tag{9}$$

Furthermore let M be a real n,n-matrix whose eigenvalues λ_i satisfy $|\lambda_i| = 1$. (10)

Then there are real values

$$\gamma_j \in [-1,+1],\ j = 1,2,\dots,n, \text{ such that } \sum_{j=1}^{n} \gamma_j M^j = \kappa(n) I. \tag{11}$$

Proof: Since M is real, the characteristic polynomial

$$\det (\lambda I - M) = \prod_{j=1}^{n} (\lambda - \lambda_j) =: \sum_{j=0}^{n} \alpha_j \lambda^j, \text{ where } \alpha_n = 1, \tag{12}$$

of M has real coefficients α_j. On the other hand by (10) we have

$$|\alpha_o| = \prod_{j=1}^{n} |\lambda_j| = 1.$$

Hence it is $\alpha_o = 1$ or $\alpha_o = -1$. (13)

If $n \geq 2$ we obtain for $j = 1,\ldots,n-1$ from Vieta's theorem

$$\alpha_j = \sum \lambda_{\nu_1} \lambda_{\nu_2} \cdots \lambda_{\nu_{n-j}} ,$$

where the summation is to be taken over the $\binom{n}{n-j}$ combinations of $n-j$ distinct indices chosen from the collection $\{1,2,\ldots,n\}$. Thus using (10) we get for $j = 1,\ldots,n-1$

$$|\alpha_j| \leq \binom{n}{n-j} = \binom{n}{j} ,$$

which by virtue of (12), (13) also holds for $j = 0$ and $j = n$. From (12) follows

$$\sum_{j=1}^{n} (-\alpha_j) M^j = \alpha_o I$$

and multiplying this by $\kappa(n) \cdot \mathrm{sign}(\alpha_o)$ we obtain (11), where for $j = 1,\ldots,n$

$$\gamma_j := -\alpha_j \, \mathrm{sign}(\alpha_o) / \max_{j=0,1,..,n} \binom{n}{j} \in [-1,+1].$$

4. Proofs of the Theorems 1 d and 1 c

Sufficiency of conditions (3d), (4d)

As it was shown in the preparatory remarks ii and iv we may restrict ourselves to the control restraint $u \in \Gamma^m_1$, that is $\|u\| \leq 1$, and assume that $\det F \neq 0$. For $k \in \mathbb{N}$ we introduce the auxiliary n,m-matrices $R_k := F^{-k}G$. Obviously we have

$$F^{-p}R_k = R_{k+p} \quad \text{for} \quad k,p \in \mathbb{N}. \tag{14}$$

The following two propositions are quite familiar in controllability theory of discrete-time systems. See for example Kalman *et.al.* [2] and Desoer and Wing [4].

Since F is regular, the controllability assumption (3d) implies

$$\operatorname{rank}\,[R_1, R_2, \ldots, R_n] = n\,. \tag{15}$$

An initial state $x^o \in \mathbb{R}^n$ at $k = 0$ can be transferred to state 0 under the control restraint $u \in \Gamma_1^m$ if and only if there is a $k_1 \in \mathbb{N}$ and a sequence

$$f_k \in \Gamma_1^m,\ k = 1,\ldots,k_1, \text{ such that } x^o = \sum_{k=1}^{k_1} R_k f_k. \tag{16}$$

In this case the control problem is solved by the admissible control $u(k) := f_{k+1}$, $k = 0, 1,\ldots,k_1 - 1$. Hence it remains to prove that for any $x^o \in \mathbb{R}^n$ a representation of type (16) is possible.

According to the preparatory remark iii we may assume that system (1d) has the form

$$\begin{bmatrix} \hat{x}(k+1) \\ \cdots\cdots \\ \tilde{x}(k+1) \end{bmatrix} = \begin{bmatrix} \hat{F} & \vdots & 0 \\ \cdots & \vdots & \cdots \\ 0 & \vdots & \tilde{F} \end{bmatrix} \begin{bmatrix} \hat{x}(k) \\ \cdots \\ \tilde{x}(k) \end{bmatrix} + G\,u(k) \tag{17}$$

where each eigenvalue $\hat{\lambda}$ of the real $\hat{n},\hat{n}$-matrix $\hat{F}$ satisfies $|\hat{\lambda}| = 1$, and each eigenvalue $\tilde{\lambda}$ of the real $\tilde{n},\tilde{n}$-matrix $\tilde{F}$ satisfies $0 < |\tilde{\lambda}| < 1$.

Let us denote the corresponding invariant subspaces by

$$\hat{L} := \{ \begin{bmatrix} \hat{x} \\ \cdots \\ \tilde{x} \end{bmatrix} \mid \tilde{x} = 0\} \quad \text{and} \quad \tilde{L} := \{ \begin{bmatrix} \hat{x} \\ \cdots \\ \tilde{x} \end{bmatrix} \mid \hat{x} = 0\} \ .$$

In order to avoid notational complications we will assume for the rest of the proof that $0 < \hat{n} < \hat{n}$. The proofs for the special cases $\hat{n} = 0$ and $\tilde{n} = 0$ will then follow immediately by cancelling all considerations referring to $\hat{L}$ or $\tilde{L}$ respectively.

First we consider $\hat{L}$. Let l be any natural number. Then each eigenvalue μ of the real $\hat{n},\hat{n}$-matrix $\hat{F}^{-1}$ satisfies $|\mu| = 1$. Hence by (10) there are real values

$$\gamma_{lj} \in [-1,+1],\ j = 1,\ldots,\hat{n}, \text{ such that } \sum_{j=1}^{\hat{n}} \gamma_{lj}\hat{F}^{-jl} = \kappa(\hat{n})I,$$

with $\kappa(\hat{n}) \in (0,1]$ as defined in (9). Therefore for each

$$x := \begin{bmatrix} \hat{x} \\ \cdots \\ 0 \end{bmatrix} \in \hat{L} \quad \text{we have} \quad \sum_{j=1}^{\hat{n}} \gamma_{lj} F^{-jl}\, x = \kappa(\hat{n})x. \tag{18}$$

Let us now take an arbitrary but fixed $x^{o} \in \hat{L}$. By virtue of (15) there is at least one set $f_1', f_2',\ldots,f_n'$ of $\mathbb{R}^m$-vectors, which satisfie

$$x^{o} = \sum_{k=1}^{n} R_k f_k'. \tag{19}$$

If $\varphi := \max_{k=1,..,n} \|f_k'\| \leq 1$ (19) is a representation of x° of the desired type (16). Otherwise we have $\varphi > 1$ and defining

$$y := \varphi^{-1} x^{\circ} = \sum_{k=1}^{n} R_k f'_k \in \hat{L}, \tag{20}$$

where $f_k := \varphi^{-1} f'_k \in \Gamma^m_1$ for $k = 1,\dots,n$, it remains to prove that for the vector
$z := x^{\circ} - y = (1 - \varphi^{-1}) x^{\circ} \in \hat{L}$ a representation

$$z = \sum_{k=n+1}^{k_1} R_k f_k$$

with a suitable $k_1 > n$ and $f_k \in \Gamma^m_1$ for $k = n + 1,\dots k_1$ can be found. Since by construction $y \neq 0$, $z \neq 0$ and $z = (\varphi - 1) y$ where $(\varphi - 1) > 0$, we can find a natural number g and a value $\alpha \in (0,1]$, such that

$$z = g \alpha \kappa(\hat{n}) y = \sum_{\nu=1}^{g} \alpha \kappa(\hat{n}) y . \tag{21}$$

At first we will construct a representation for a single term $\alpha \kappa(\hat{n}) y$. For any fixed natural number $l \geq n$ and for $j = 1,2,\dots,\hat{n}$ we apply $\alpha \gamma_{lj} F^{-jl}$ to equation (2) from the left hand side. Thus, using (14), we obtain the $\hat{n}$ equations

$$\alpha \gamma_{lj} F^{-jl} y = \sum_{k=1}^{n} R_{k+jl} \alpha \gamma_{lj} f_k , \quad j = 1,2,\dots,\hat{n}. \tag{22}$$

Since $l + 1 > n$ we have $n + jl < 1 + (j+1)l$ for $j = 1,2,\dots,\hat{n} - 1$. Thus on the right hand side of system (22) no index $k + jl$ is repeated. From $\alpha \in (0,1]$, $\gamma_{lj} \in [-1,+1]$ and $f_k \in \Gamma^m_1$ it is clear that for all indices $\mu := k + jl$ appearing in system (22), $f_\mu := \alpha\gamma_{lj} f_k \in \Gamma_1^m$ holds. Setting $f_\mu := 0 \in \Gamma_1^m$ for such indices $\mu \in \{l + 1,...,n + \hat{n}l\}$ which do not appear in (22) and summing up the $\hat{n}$ equations in (22) we get

$$\alpha\kappa(\hat{n})y = \sum_{\mu=l+1}^{n+\hat{n}l} R_\mu f_\mu, \text{ with } f_\mu \in \Gamma_1^m \text{ for } \mu = l+1,..,n+\hat{n}l, \tag{23}$$

where we have used (18) for the summation on the left hand side. Let us now define $l_1 := n$ and recursively $l_\nu := n + \hat{n}l_{\nu-1}$ for $\nu = 2,...,g + 1$. For each $\nu \in \{1,...,g\}$ we can express $\alpha\kappa(\hat{n})y$ by a sum of type (23), where the summation runs from $\mu = l_\nu + 1$ to $\mu = l_{\nu+1}$. Substitution of these nonoverlapping representations of $\alpha\kappa(\hat{n})y$ into (21) yields

$$z = \sum_{\nu=1}^{g} \sum_{\mu=l_\nu+1}^{l_{\nu+1}} R_\mu f_\mu$$

which is obviously a representation of z of the type desired.

Next we consider an arbitrary but fixed initial state

$$x^0 := \begin{bmatrix} 0 \\ \cdots \\ \tilde{x}^0 \end{bmatrix} \in \tilde{L} .$$

From $\tilde{F}^q \longrightarrow 0$ for $q \to \infty$ it follows $\|F^q x^0\| \longrightarrow 0$ for $q \to \infty$. Therefore and on account of (15) we can find such $p \in \mathbb{N}$ that the linear system

$$F^p x^o = \sum_{j=1}^{n} R_j f'_j$$

admits a solution $f'_1, f'_2, \ldots, f'_n \in \Gamma_1^m$. Thus, using (14) and defining $f_k := 0 \in \Gamma_1^m$ for $k = 1, \ldots, p$ and $f_k := f'_{k-p} \in \Gamma_1^m$ for $k = p+1, \ldots, p+n$ we can write

$$x^o = \sum_{k=1}^{n+p} R_k f_k$$

which is the representation of x^o we wanted.

Finally we consider an arbitrary initial state $x^o = x^{o1} + x^{o2}$ with $x^{o1} \in \hat{L}$ and $x^{o2} \in \tilde{L}$. By the two preceeding results there are representations

$$2x^{o1} = \sum_{k=1}^{k_1} R_k f_k^1 \quad \text{and} \quad 2x^{o2} = \sum_{k=1}^{k_1} R_k f_k^2$$

with an appropriate $k_1 \in \mathbb{N}$ and $f_k^{1,2} \in \Gamma_1^m$ for $k = 1, 2, \ldots k_1$. Here we used the fact that in case of need a given representation can be extended up to summation limit k_1 by adding 0-terms. Summing up the two representations above yields

$$x^o = \sum_{k=1}^{k_1} R_k \tfrac{1}{2} (f_k^1 + f_k^2),$$

where obviously $\tfrac{1}{2}(f_k^1 + f_k^2) \in \Gamma_1^m$ for $k = 1, \ldots, k_1$.

□

Sufficiency of conditions (3c), (4c)

On account of well-known results concerning the preservation of controllability under sampling, sufficiency of (3c), (4c) turns out to be an easy consequence of the sufficiency of conditions (3d), (4d) just proved.

Let $T > 0$ be arbitrary for the present and apply for any given sequence $u(k) \in \mathbb{R}^m$, $k = 0,1,\ldots$, to system (1c) the control $v(\cdot)$ defined by

$$v(t) := u(k) \quad \text{for} \quad kT \leq t < (k+1)T, \; k = 0,1,\ldots \; . \tag{24}$$

Restricting our attention to the sampling instants $t = kT$ and defining $x(k) := y(kT)$ we obtain from (1c)

$$x(k+1) = F\,x(k) + G\,u(k)\,, \; k = 0,1,2,\ldots \quad , \tag{25}$$

where $F := e^{TA}$ and $G := \int_0^T e^{(T-\tau)A} \, B \, d\tau$.

Let us now assume that the system (1c) satisfies the conditions (3c) and (4c). Then for every choice of $T > 0$ (4c) implies condition (4d) with respect to system (25). Furthermore, as it is well-known, (e. g. Kalman *et al.* [2], Hautus [10], Bar-Ness and Langholz [12]) one can choose an appropriate $T > 0$ such that the controllability condition (3c) implies the controllability condition (3d) with respect to system (25). Thus for such a choice of T we may apply to system (25) the sufficiency part of theorem 1d, which is already proved: For every $y^0 =: x^0 \in \mathbb{R}^n$ we can find an integer $k_1 > 0$ and a control $u(\cdot) : \{0,1,..,k_1 - 1\} \longrightarrow \Gamma_1^m$ for system (25) such that $x(k_1;0, x^0,u(\cdot)) = 0$. Hence we have for system (1c) a control $v(\cdot) : [0,k_1T) \longrightarrow \Gamma_1^m$, defined by (24), such that $y(k_1T;\, 0,\, y^0,\, v(\cdot)) = 0$.

Here we proved the sufficiency of conditions (3c), (4c)

by constructing a piecewise constant control $v(\cdot)$, which of course will not be smooth in general, whereas Lee and Markus [7] in their proof for the special case $m = 1$ constructed a smooth control. Therefore we will sketch in the sequel a modification of the above proof showing the existence of a smooth admissible control in the general case $m \geq 1$ as well. Let us keep the $T(> 0)$ choosen above which assured the controllability of system (25). For $\delta \in (0, T/4)$, arbitrary for the present, take a C^{∞}-function $s_{\delta}(\cdot) : [0,T] \longrightarrow \mathbb{R}$, as it is frequently used as test function in distribution theory, with the following properties

i $\quad s_{\delta}(\tau) = 0$ for $0 \leq \tau \leq \delta$ and for $T-\delta \leq \tau \leq T$

ii $\quad 0 \leq s_{\delta}(\tau) \leq 1$ for $\delta \leq \tau \leq 2\delta$ and for $T-2\delta \leq \tau \leq T-\delta$

iii $\quad s_{\delta}(\tau) = 1$ for $2\delta \leq \tau \leq T-2\delta$.

Replace the sampling mechanism (24) by

$$v(t) := s_{\delta}(t-kT)u(k) \text{ for } kT \leq t \leq (k+1)T\ , \ k = 0,1,\ldots \qquad (\tilde{24})$$

Obviously this mechanism produces C^{∞}-controls only.

Instead of (25) we obtain now

$$x(k+1) = F\,x(k) + \tilde{G}_{\delta}\, u(k)\ , \ k = 0,1,2,\ldots, \qquad (\tilde{25})$$

with the same F as in (25) and G replaced by

$$\tilde{G}_{\delta} := \int_{0}^{T} s_{\delta}(\tau)\, e^{(T-\tau)A}\, B\, d\tau\ .$$

But for any given $\varepsilon > 0$ we can find a $\delta_{o} > 0$ such that $\|G - \tilde{G}_{\delta}\| < \varepsilon$ if $\delta \in (0, \min\{T/4, \delta_{o}\})$.

Hence the null-controllability of system (25) implies the null-controllability of system $(\tilde{25})$ for sufficiently small $\delta > 0$. In this conclusion we used the fact that controllability properties of time-invariant systems are generic (e. g. W. M. Wonham [11]). The further procedure then is the same as in the previously treated case of constant sampling.

Necessity of conditions 3d, 3c and 4c

Since null-controllability under any control restraint always implies null-controllability in the absence of control restraints, the necessity of conditions 3d and 3c is evident. Lee and Markus [7] gave a proof for the necessity of condition 4c in the special case $m = 1$ which after a slight notational modification also applies if $m \geq 1$. It will not be repeated here. Moreover in the next section the idea of this proof will be transferred to the discrete-time case.

Necessity of condition 4d

For an indirect proof we suppose that F has an eigenvalue. λ_1 with $|\lambda_1| > 1$. We may assume that F is given in a real Jordan canonical form. Two cases arise.

i. λ_1 is real and $|\lambda_1| = 1 + \varepsilon$ with an $\varepsilon > 0$. At least one line of system (1d) is in the form

$$x_\nu(k+1) = \lambda_1 x_\nu(k) + b_\nu u(k) \, , \tag{26}$$

where x_ν is the ν-th component of the state vector x and b_ν ist the ν-th line of the matrix B. Since Ω is bounded we can find an appropriate $\gamma > 0$ such that for any admissible control $u(k) \in \Omega$ holds

$|b_\nu u(k)| \leq \gamma$. Thus we obtain from (26)

$$|x_\nu(k+1)| \geq |x_\nu(k)| + \varepsilon |x_\nu(k)| - \gamma \tag{27}$$

as long as admissible controls are applied.

If $|x_\nu(0)| \geq \gamma/\varepsilon$ it follows from (27) by induction that $|x_\nu(k)| \geq \gamma/\varepsilon > 0$ for $k = 0,1,\ldots$. This means that an initial state $x(0) \in \mathbb{R}^n$ with $x_\nu(0) \geq \gamma/\varepsilon$ never can be steered to 0 under the control restraint $u \in \Omega$.

ii. $\gamma = \alpha + i\beta$, $\alpha,\beta \in \mathbb{R}$, is not real and $\alpha^2 + \beta^2 = 1 + \varepsilon$ with an $\varepsilon > 0$.

At least two consecutive lines of system (1d) have the form

$$x_\nu(k+1) = \alpha x_\nu(k) - \beta x_{\nu+1}(k) + b_\nu u(k)$$

$$x_{\nu+1}(k+1) = \beta x_\nu(k) + \alpha x_{\nu+1}(k) + b_{\nu+1} u(k).$$

An easy calculation yields

$$\begin{aligned}[x_\nu^2(k+1) + x_{\nu+1}^2(k+1)] &= (1+\varepsilon)[x_\nu^2(k) + x_{\nu+1}^2(k)] + \\ &+ (b_\nu u(k))^2 + (b_{\nu+1}u(k))^2 + \\ &+ x_\nu(k)(2\alpha b_\nu + 2\beta b_{\nu+1})u(k) + x_{\nu+1}(k)(2\alpha b_{\nu+1} - 2\beta b_\nu)u(k). \qquad (28)\end{aligned}$$

Obviously we can find a $\gamma > 0$ such that for all $u(k) \in \Omega$

$$\{((2\alpha b_\nu + 2\beta b_{\nu+1})u(k))^2 + ((2\alpha b_{\nu+1} - 2\beta b_\nu)u(k))^2\}^{\frac{1}{2}} \leq \gamma .$$

Thus we obtain from (28) by means of the Cauchy-Schwarz inequality

$$\begin{aligned}[x_\nu^2(k+1) + x_{\nu+1}^2(k)] \geq\ & (1+\varepsilon)[x_\nu^2(k) + x_{\nu+1}^2(k)] \\ & - \gamma[x_\nu^2(k) + x_{\nu+1}^2(k)]^{\frac{1}{2}} \qquad (29)\end{aligned}$$

as long as admissible controls are applied.

If $[x_\nu^2(0) + x_{\nu+1}^2(0)] \geq (\gamma/\varepsilon)^2 > 0$ it follows from (29) by induction that $[x_\nu^2(k) + x_{\nu+1}^2(k)] \geq (\gamma/\varepsilon)^2 > 0$ for $k = 0,1,\ldots$. This shows that no initial state $x(0) \in \mathbb{R}^n$ with $[x_\nu^2(0) + x_{\nu+1}^2(0)] \geq (\gamma/\varepsilon)^2$ can be brought to the origin by application of an admissible control.

References

[1] R.E. Kalman
On the General Theory of Control Systems,
Proceedings of the first International Congress on Automatic Control, Moscow, 1960,
Butterworths Scientific Publications, Vol. 1, London 1961, p. 481.

[2] R.E. Kalman, Y.C. Ho, K.S. Narendra
Controllability of Linear Dynamical Systems,
Contr. Diff. Eqns., Vol. 1, 1963, p. 189.

[3] L.S. Pontryagin, V.G. Boltyanskii, R.V. Gamkrelidze, E.F. Mishchenko
The Mathematical Theory of Optimal Processes,
Interscience Publishers, New York 1962.
Translation from the Russian original, Moscow 1961.

[4] C.A. Desoer, J. Wing
The Minimal Time Regulator Problem for Linear Sampled-Data Systems: General Theory,
Journal of the Franklin Institute, Vol.272, 1961, p.208.

[5] M. Athans, P.L. Falb
Optimal Control,
McGraw-Hill, New York 1966.

[6] R.W. Koepke
A Solution to the Sampled, Minimum-Time Problem,
Journal of Basic Engineering, Vol. 86, 1964, p. 145.

[7] E.B. Lee, L. Markus
Foundations of Optimal Control Theory,
John Wiley & Sons, New York 1967.

[8] H.-J. Kowalski
Lineare Algebra,
Walter de Gruyter & Co., Berlin 1967.

[9] G. Ludyk
Zeitoptimale Abtastsysteme mit Beschränkung der Stellgröße,
Regelungstechnik, Vol. 16, 1968, p. 344.

[10] M.L.J. Hautus
Controllability and Stabilizability of Sampled Systems
IEEE Trans. on Aut. Contr., Vol. AC 17, 1972, p. 528.

[11] W.M. Wonham
Linear Multivariable Control,
Springer, Berlin, Heidelberg 1974.

[12] Y. Bar-Ness, G. Langholz
Preservation of Controllability under Sampling,
Int. Journal of Control, Vol. 22, 1975, p. 39.